D1037453

BIOMECHANICS AT MICRO- AND NANOSCALE LEVELS

VOLUME III

Biomechanics at Micro- and Nanoscale Levels

Editor-in-Charge: Hiroshi Wada
(Tohoku University, Sendai, Japan)

Published

Vol. I: Biomechanics at Micro- and Nanoscale Levels
Edited by Hiroshi Wada
ISBN 981-256-098-X

Vol. II: Biomechanics at Micro- and Nanoscale Levels
Edited by Hiroshi Wada
ISBN 981-256-746-1

BIOMECHANICS AT MICRO- AND NANOSCALE LEVELS

Selected Papers from the Proceedings of the Fifth World Congress of Biomechanics, Thread 3: Biomechanics at Micro- and Nanoscale Levels

VOLUME III

editor

Hiroshi Wada

Tohoku University, Sendai, Japan

NEW JERSEY • LONDON • SINGAPORE • BEIJING • SHANGHAI • HONG KONG • TAIPEI • CHENNAI

Published by

World Scientific Publishing Co. Pte. Ltd.

5 Toh Tuck Link, Singapore 596224

USA office: 27 Warren Street, Suite 401-402, Hackensack, NJ 07601

UK office: 57 Shelton Street, Covent Garden, London WC2H 9HE

British Library Cataloguing-in-Publication Data
A catalogue record for this book is available from the British Library.

BIOMECHANICS AT MICRO- AND NANOSCALE LEVELS
Volume III

ISBN-13 978-981-270-814-4
ISBN-10 981-270-814-6

Printed by Mainland Press Pte Ltd

PREFACE

A project on "Biomechanics at Micro- and Nanoscale Levels," the title of this book, was approved by the Ministry of Education, Culture, Sports, Science and Technology of Japan in 2003, and this four-year-project is now being carried out by fourteen prominent Japanese researchers. The project consists of four fields of research, namely, Cell Mechanics, Cell Response to Mechanical Stimulation, Tissue Engineering, and Computational Biomechanics.

Our project can be summarized as follows. The essential diversity of phenomena in living organisms is controlled not by genes but rather by the interaction between the micro- or nanoscale structures in cells and the genetic code, the dynamic interaction between them being especially important. Therefore, if the relationship between the dynamic environment of cells and tissues and their function can be elucidated, it is highly possible to find a method by which the structure and function of such cells and tissues can be regulated. The first goal of this research is to understand dynamic phenomena at cellular and biopolymer-organelle levels on the basis of mechanics. An attempt will then be made to apply this understanding to the development of procedures for designing and producing artificial materials and technology for producing or regenerating the structure and function of living organisms.

At the 5th World Congress of Biomechanics held in Munich, Germany from 29th July to 4th August, 2006, we organized the following sessions:

Thread 3: Biomechanics at micro- and nanoscale levels

1. Cell mechanics
2. Molecular biomechanics
3. Mechanobiology at micro- and nanoscale levels
4. Computational biomechanics

We are planning to publish a series of books related to this project, the present volume being proceedings covering topics related to the sessions we organized. I would like to express my sincere gratitude to Professor Dieter Liepsch, the President of the 5th World Congress of Biomechanics, who granted us permission to publish these proceedings, as well as to the ten researchers who contributed to these proceedings.

Hiroshi Wada, PhD,
Project Leader,
Tohoku University,
Sendai,
December, 2006

5th World Congress of Biomechanics

FOREWORD

Munich, December, 2006

This volume represents the finest research of some of the best investigators in the world and I am very honored that these important and ground-breaking papers were presented at the 5th World Congress of Biomechanics in Munich, in 2006.

Between July 31st and August 4th, 2006, nearly 2600 scientists and 900 students from 64 countries, representing every continent, came to Munich, Germany to listen, learn and share their latest work in all fields of biomechanics. It quickly became obvious the newest trends lay in the fields of tissue and cellular biomechanics. It was also clear that the field was so rich and multi-faceted that it warranted an orthogonal 'thread' which would cut across all the various fields of research, showing the connections, interrelations and interdependencies. Under the guidance of Prof. Dr. Hiroshi Wada, Thread 3 "Biomechanics at micro- and nanoscale levels" incorporated cell mechanics, molecular biomechanics, mechanobiology at micro- and nano-scale levels and computational biomechanics.

When Prof. Hiroshi Wada first raised the possibility of incorporating micro- and nanoscale biomechanics into the 5th World Congress of Biomechanics, I knew that without this contribution, we could not hope to present to the international community the full scope and richness of this increasingly vital field of research. It was my pleasure and honor to be able to present to the international scientific world and to our students, the work of the scientists represented in this volume. This volume is the work of dedicated scientists who have earned the respect and esteem of their peers around the world and who have opened the door to new vistas of research for a new generation of students and young investigators. I am grateful to have played a small part in bringing this work to the international scientific community.

With best regards

Dieter Liepsch
Congress President

CONTENTS

PREFACE v

FOREWORD vii

I. CELL MECHANICS 1

The effect of streptomycin and gentamicin on outer hair cell motility 3
B. Currall, X. Wang and D. Z. Z. He

Mechanotransduction in bone cell networks 13
X. E. Guo, E. Takai, X. Jiang, Q. Xu, G. M. Whitesides, J. T. Yardley, C. T. Hung and K. D. Costa

Intracellular measurements of strain transfer with texture correlation 36
C. L. Gilchrist, F. Guilak and L. A. Setton

II. CELL RESPONSE TO MECHANICAL STIMULATION 49

Identifying the mechanisms of flow-enhanced cell adhesion via dimensional analysis 51
C. Zhu, V. I. Zarnitsyna, T. Yago and R. P. McEver

A sliding-rebinding mechanism for catch bonds 66
J. Lou, C. Zhu, T. Yago and R. P. McEver

Role of external mechanical forces in cell signal transduction 80
S. R. K. Vedula, C. T. Lim, T. S. Lim, G. Rajagopal, W. Hunziker, B. Lane and M. Sokabe

III. TISSUE ENGINEERING 105

Evaluation of material property of tissue-engineered cartilage by magnetic resonance imaging and spectroscopy 107
S. Miyata, K. Homma, T. Numano, K. Furukawa, T. Ushida and T. Tateishi

Scaffolding technology for cartilage and osteochondral tissue engineering 118
G. Chen, N. Kawazoe, T. Tateishi and T. Ushida

IV. COMPUTATIONAL BIOMECHANICS 131

MRI measurements and CFD analysis of hemodynamics in the aorta and the left ventricle 133
M. Nakamura, S. Wada, S. Yokosawa and T. Yamaguchi

A fluid-solid interactions study of the pulse wave velocity in uniform arteries 146
T. Fukui, Y. Imai, K. Tsubota, T. Ishikawa, S. Wada, T. Yamaguchi and K. H. Parker

Rule-based simulation of arterial wall thickening induced by low wall shear stress 157
S. Wada, M. Nakamura and T. Karino

SUBJECT INDEX 167

I. CELL MECHANICS

THE EFFECT OF STREPTOMYCIN AND GENTAMICIN ON OUTER HAIR CELL MOTILITY

B. CURRALL, X. WANG AND D. Z. Z. HE

Department of Biomedical Sciences, Creighton University School of Medicine
Omaha, Nebraska, 68178, USA
E-mail: hed@creighton.edu

The cochlear outer hair cell (OHC), which plays a crucial role in mammalian hearing through its unique voltage-dependent length change, has been established as a primary target of the ototoxic activity of aminoglycoside antibiotics. Although the ototoxicity eventually leads to hair cell loss, these polycationic drugs are also known to block a wide variety of ion channels such as mechanotransducer channels, purinergic ionotropic channels and nicotinic ACh receptors in acute preparations. The OHC motor protein prestin is a voltage-sensitive transmembrane protein which contains several negatively charged residues on both intra- and extracellular surface. The acidic sites suggest that they may be susceptible to polycationic-charged aminoglycoside binding, which could result in a disruption of somatic motility. We attempted to examine whether aminoglycosides such as streptomycin and gentamicin could affect the mechanical response of OHCs. Solitary OHCs isolated from adult gerbils were used for the experiments. Somatic motility and nonlinear capacitance were measured under the whole-cell voltage-clamp mode. Streptomycin and gentamicin were applied extracellularly or intracellularly. Results show that streptomycin and gentamicin, for the concentration range between 100 μM and 1 mM, did not affect somatic motility or nonlinear capacitance. The result suggests that although streptomycin and gentamicin can block mechanotransduction channels as well as ACh receptors in hair cells, they have no immediate effect on OHC somatic motility.

1 Introduction

Aminoglycosides are low cost, high efficacy antibiotics, however, their use is limited by their nephrotoxic and ototoxic activity. Several toxic mechanisms have been associated with aminoglycosides. In genetically susceptible individuals, a mitchodonrial mutation for an rRNA may be vulnerable to aminoglycoside interference [1]. It has also been shown that *N*-methyl-D-apartate (NMDA) receptor, found in afferent neurons, may be affected by aminoglycosides, resulting in excitotoxicity followed by hair cell death [2]. Also, upon entry into the cell, whether via vesicle-mediated process [3] or the mechanoelectical transuction channel [4], reactive oxygen species, free radicals and nitiric oxide form, resulting in multiple signaling pathways that may lead to subsequent cell death [5-8]. While nephro- and ototoxicity seem to depend on intracellular accumulation of these antibiotics [9-10], numerous studies have demonstrated the ability of these polycationic drugs to acutely depress synaptic transmission at the neuromuscular junction, presumably by blocking presynaptic voltage-gated Ca^{2+} channels [11-12]. These polycationic drugs also block a wide variety of ion channels such as mechanosensitive ion channels [13-14], purinergic ionotropic channels [15] and nicotinic ACh receptors [16-17].

In the cochlea, the ototoxic activity of aminoglycosides is characterized by a loss of outer hair cells (OHCs). The OHC is one of two receptor cells in the organ of Corti, and plays a critical role in mammalian hearing. OHCs are able to rapidly change their length [18-19] and stiffness [20] at acoustic frequencies when their transmembrane potential is altered. This fast somatic motility is believed to be the substrate of cochlear amplification [18, 21]. OHC electromotility is driven by voltage-sensitive molecules (or assemblies of molecules) able to change area when the membrane potential is altered. Both the motor and its sensor are located in the plasma membrane [22-24]. Recently, the gene *prestin* that codes the motor protein was identified [25]. The targeted deletion of prestin in mice results in the loss of electromotility *in vitro*, and a 40-60 dB loss in cochlear sensitivity *in vivo* [21]. The protein prestin is a voltage-sensitive transmembrane protein which contains several negatively charged residues on both intra- and extracellular surface [26]. The acidic sites suggest that they may be susceptible to aminoglycoside binding [27], which could result in a disruption of somatic motility. Therefore, the purpose of this study was to determine whether the aminoglycosides such as streptomycin and gentamicin could affect the mechanical response of OHCs.

2 Materials and Methods

2.1 Preparation of isolated OHCs

Gerbils (*Meriones unguiculatus*) ranging in age between 4 and 8 weeks were anesthetized with an intraperitoneal injection of a lethal dose of sodium pentobarbital (150 mg/kg) and then decapitated. Cochleae were dissected out and kept in cold culture medium (Leibovitz's L-15). L-15 (Gibco) was supplemented with 10 mM HEPES (Sigma) and adjusted to 300 mOsm and pH 7.4. After the cochlear wall was removed, the BM-organ of Corti complex was unwrapped from the modiolus from the base to the apex. The organ of Corti was dissected out from the apical turn of the cochlea. The tissue was then transferred to the enzymatic digestion medium [L-15 supplemented with 1 mg/ml collagenase type IV (Sigma)]. After 10 minutes incubation at room temperature (22±2°C), the tissue was transferred to the experimental bath containing fresh L-15 medium. To obtain solitary OHCs, gentle trituration of the tissue with a small pipette was applied. A cell was selected for experimentation only if its diameter was approximately constant throughout its length and if it showed no signs of damage, such as swelling, blebbing, and dislocation of the nucleus. Cells were rejected if visible signs of damage and appearance changes occurred during the experiment.

2.2 Whole-cell voltage-clamp recording

Isolated OHCs were placed in the experimental chamber containing extracellular fluid (pH 7.2, 320 mOsm, in mMol: $NaCl_2$ 120, TEA-Cl 20, $CoCl_2$ 2.0, $MgCl_2$ 2.0, $CaCl_2$ 1.5, $HEPES_2$ 10, Glucose 5.0) on the stage of an inverted microscope (Olympus IX–71). The patch electrodes were pulled from 1.5 mm glass capillaries (A-M System) using a Flaming/Brown Micropipette Puller (Sutter Instrument Company, Model P-97). The electrodes were back-filled with solution containing (in mM) CsCl 140; $CaCl_2$; 0.1; $MgCl_2$ 3.5; Na_2ATP; 2.5; EGTA-KOH 5; HEPES-KOH 10. The solution was adjusted to pH 7.4 with CsOH (Sigma) and osmolarity adjusted to 300 mOsm with glucose. These solutions enabled to block K^+ and Ca^{2+} conductances to isolate gating currents associated with somatic motility. The pipettes had initial bath resistances of 2-4 MΩ. The access resistance, that is, the actual electrode resistance obtained upon establishment of the whole-cell configuration, typically ranged from 6 to 12 MΩ. Series resistance was corrected off-line after data collection. Aminoglycosides were applied either extracellularly, with aminoglycosides placed in separate perfusion pipette, or intracellularly, with aminoglycosides placed in the patch pipette. Extracellular perfusion pipettes were placed within 60 μm of cell, after achieving whole-cell configuration, and perfusion started by opening gravity fed T-tubule switch. Streptomycin sulfate, and gentamicin sulfate were diluted to concentration specified in figures captions.

2.3 Somatic motility measurements

Somatic motility was measured and calibrated by photodiode-based measurement systems [28] mounted on the Olympus inverted microscope. The OHC was imaged using a 40x objective and magnified by an additional 20x relay lens. The magnified image of the edge of the cell was then split into two paths: one path projected onto the photodiode (Hamamatsu) through a slit and another projected onto a CCD camera so that the edge of the cell could be viewed at all times on a television monitor. During measurements, the magnified image of the edge of the cell was positioned near the edge of the slit. The slit was rotated, based on the orientation of the cell. The photodiode system had a cutoff (3-dB) frequency of 1,200 Hz. The signal was then amplified by a 60-dB fixed-gain dc-coupled amplifier. The amplified signal was then low-pass filtered (400 or 1,100 Hz) before being delivered to one of the A/D inputs of a Digidata (1322A, Axon Instruments) acquisition board in a Window-based PC. The measurement system was capable of measuring motions down to ~5 nm with 100 averages. Calibration was performed by moving the slit a known distance (1 μm).

2.4 Nonlinear capacitance measurements

The AC technique was used to obtain motility-related gating charge movement and the corresponding NLC. This technique has been described in details elsewhere [29]. In brief, it utilized a continuous high-resolution (2.56 ms sampling) two-sine voltage stimulus protocol (10 mV peak at both 390.6 and 781.2 Hz), with subsequent fast Fourier transform-based admittance analysis. These high frequency sinusoids were superimposed on voltage ramp stimuli. The NLC can be described as the first derivative of a two-state Boltzmann function relating nonlinear charge movement to voltage [30-31]. The capacitance function is described as:

$$C_m = \frac{Q_{\max}\alpha}{\exp[\alpha(V_m - V_{1/2})](1+\exp[-\alpha(V_m - V_{1/2})])^2} + C_{lin}$$

where, Q_{max} is maximum charge transfer, $V_{1/2}$ is the voltage at which the maximum charge is equally distributed across the membrane, C_{lin} is linear capacitance, and $\alpha = ze/kT$ is the slope factor of the voltage dependence of charge transfer where k is Boltzmann's constant, T is absolute temperature, z is valence, and e is electron charge. Capacitive currents were filtered at 2 kHz and digitized at 10 kHz using jClamp software (SciSoft Company), running on an IBM-compatible computer and a 16-bit A/D converter (Digidata 1322A, Axon Instruments).

3 Results

3.1 Extracellular application of streptomycin and gentamicin

OHC motility was measured from isolated cells before and after streptomycin and gentamicin were applied to the extracellular solution through a puffer pipette positioned 60 µm away from the cells. The cells were held at -70 mV and voltage steps varying from -120 mV to 60 mV were used to evoke motility. Fig. 1 shows examples of two OHCs before and 2-minutes after 100 µM streptomycin and gentamicin were applied. The motile response was asymmetric, with contraction being larger than the elongation. The response was also nonlinear, with saturation at both directions. We measured a total of 10 cells (5 cells each) for streptomycin and gentamicin at the concentration of 100 µM. Streptomycin and gentamicin did not change the magnitude nor the asymmetry of the response as shown in Fig. 1.

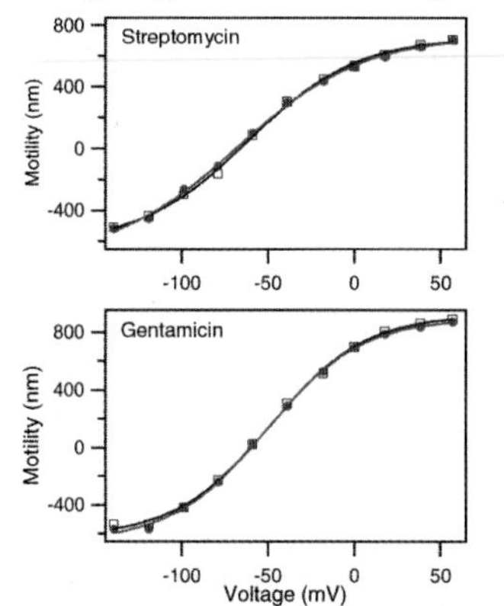

Figure 1. Voltage to length change function measured before (in black) and 2-minutes after 100 µM streptomycin and gentamicin were applied (in red). The cells were held at -70 mV under whole-cell voltage-clamp mode. Voltage steps varying from -120 mV to 60 mV were applied to evoke motility. Motility magnitude was measured from the steady-state responses. Voltage error due to series resistance was compensated. Note that neither the magnitude nor the response characteristics were changed after the treatment.

Associated with the OHC electromotility is an electrical signature, a voltage-dependent capacitance or, correspondingly, a gating charge movement [30-31], similar to the gating currents of voltage-gated ion channels [32]. The gating currents are thought to arise from a redistribution of charged voltage sensors across the membrane. This charge movement imparts a bell-shaped voltage dependence to the membrane capacitance [30-31]. Measures of nonlinear capacitance (NLC) have been used to assay OHC's motor function [25, 31]. We measured NLC before and after streptomycin and gentamicin were applied to the cells. Fig. 2 shows some representative responses from two OHCs. NLCs were measured before and 2-minutes after 100 μM streptomycin and gentamicin were applied through a perfusion pipette positioned 60 μm away from the cells. As shown, the capacitance function was bell-shaped with respect to stimulating voltage. The NLC exhibited a peak around -50 mV for both cells showed in the example. As shown, the magnitude of the peak capacitance did not change significantly after the treatment. We compared the maximum charge transfer (Q_{max}), $C_{non\text{-}lin}$, and slope factor (α). None of them changed significantly after the treatment.

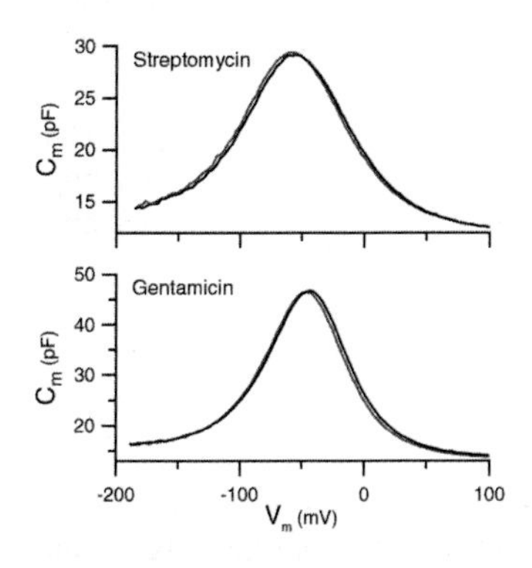

Figure 2. Capacitance measured from two OHCs before (in balck) and 2-miuntes (in red) after 100 μM streptomycin and gentamicin were applied to the extracellular solution. Note that neither the magnitude of the peak capacitance nor the $V_{1/2}$ changed after perfusion.

The changes in peak capacitance (Cm_{pk}) have been used to examine the effects of certain agents on OHC motility [33]. We also monitored Cm_{pk} during application of streptomycin and gentamicin to further determine their influence on motility. For positive control, we monitored the change in Cm_{pk} after salicylate was applied extracellularly. Salicylate is known to significantly reduce OHC somatic motility and NLC [26, 33]. Cm_{pk} was monitored using the software in the jClamp (version 12.1) package over the course of 2 to 4 minutes after aminoglycosides or salicylate was applied. Fig. 3 shows an example of such recordings. As shown, 5 mM salicylate caused significantly reduction in Cm_{pk}. However, neither streptomycin nor gentamicin had any affect on Cm_{pk}.

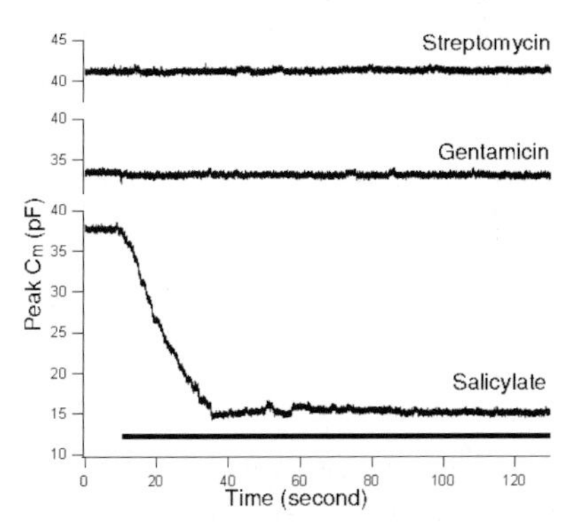

Figure 3. Peak capacitance (Cm_{pk}) monitored after streptomycin and gentamicin was applied to the extracellular solution. For positive control, 5 mM salicylate was applied. The cells were held at -40 mV under whole-cell voltage-clamp condition. Cm_{pk} was measured using the software in the jClamp package. Bar indicates the duration that streptomycin was perfused. Note that Cm_{pk} was significantly reduced after 5 mM salicylate was applied. However, neither streptomycin nor gentamicin had any affect on Cm_{pk}.

While it has been demonstrated that 100 μM streptomycin or gentamicin is enough to block mechanoelectrical transducer current as well as ACh receptor in hair cells, we inquired whether it requires even higher concentration to be effective to affect OHC somatic motility. Peak capacitance was monitored using the software in the jClamp package over the course of experiments when streptomycin with concentration of 0.1, 0.2, 0.5 and 1 mM was applied. Fig. 4 illustrates the peak capacitance measured from OHCs in response to different concentrations of streptomycin applied to the extracellular solution. As shown, the peak capacitance did not change for all the concentrations applied.

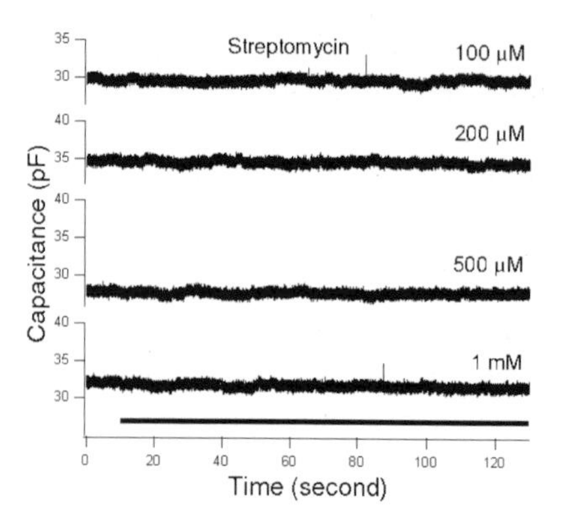

Figure 4. Peak capacitance monitored with different concentration of streptomycin applied to the extracellular solution. Peak capacitance was measured using the software in the jClamp package. Bar indicates the duration when streptomycin was perfused.

3.2 Intracellular application of streptomycin

Aminoglycosides can enter hair cells through the open mechanotransducer channels [4]. So we inquired whether aminoglycosides disturbed somatic motility when they were applied intracellularly. We examined such possibility by monitoring somatic motility immediately after rupturing the cells and throughout the entire course when the streptomycin (together with normal intracellular solution) in the patch electrode diffused to the cytosol of the cells. Fig. 5 illustrates an example of such recordings from a gerbil apical turn OHC. The cell was held at -40 mV and a 5 Hz sinusoidal voltage command with peak-to-peak amplitude of 30 mV was continuously applied to the cell to evoke motility. Motility was measured using a photodiode-based displacement system. Since the streptomycin in the patch electrode diffused to the cytosol took time (normally it would take 20 to 30 second to equilibrate), the motility measured immediately after rupturing was used as control [34]. Motility

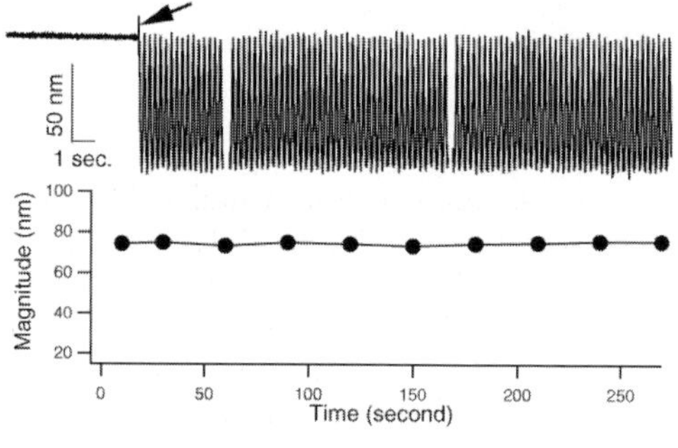

Figure 5. Motility measured after streptomycin was applied intracellularly through the patch electrode. Arrow indicates the moment when the cell's membrane was ruptured and streptomycin started to diffuse to the cytosol of the cell. The cell was held at -40 mV and 5 Hz sinusoidal voltage stimulus with peak-to-peak magnitude of 30 mV was continuously delivered to the cell to evoke motility. Three representative responses in the top panels were acquired at 10 seconds before and after the cell was ruptured, and at 30 and 200 seconds after the cell was ruptured. Steady-state responses (peak-to-peak) at different moments during perfusion were measured and plotted in the bottom panel.

was observable immediately after the cell was ruptured. We measured the magnitude of motility at different times during perfusion and the magnitude of motility is plotted in the bottom panel of Fig. 5. As shown, the magnitude of motility remained basically the same throughout the course of equilibrium. This suggests that motility is not affected by intracellular application of streptomycin.

4 Discussion

Aminoglycosides are large, lipid insoluble, polycationic molecules that are known to block a variety of ion channels including large-conductance Ca^{2+}-activated K^+ channels [35], Ca^{2+} channels [36] and ryanodine receptors [37]. In hair cells, aminoglycosides have been reported to block transducer channels [13-14], ATP receptors [15], nicotinic acetylcholine receptors [16] and large-conductance Ca^{2+}-activated K^+ channels [16]. Aminoglycosides have a strong propensity to associate with negatively charged lipid bilayers [38] and to compete at Ca^{2+} binding sites on the plasma membrane of OHCs [39].

Contrary to our expectations that streptomycin or gentamicin would be able to screen a significant proportion of fixed negative charges in prestin, we saw no reduction in either NLC or somatic motility by gentamicin or streptomycin. Though 100 μM of either streptomycin or gentamicin has been found potent enough to block mechanotransducer channels and ACh receptors (cite), we found no influence on somatic motility despite concentrations as high as 1 mM. We monitored the change in peak capacitance for over 4 minutes, long enough to see the effect if any. It is possible that such screening effect on negative charges do not affect the function of prestin.

Aminoglycosides enter hair cells via vesicle-mediated process [3] or the mechanoelectrical transuction channel [4]. Since the hair bundle is usually damaged in isolated OHCs, it is possible that the concentration of streptomycin or gentamicin inside the cell was too low to produce any effect. However, since we also did not see any effect when streptomycin was applied intracellularly, such possibility could be rule out.

Despite the high concentrations used, this study does not eliminate the possibility that aminoglycosides may have some effect on somatic motility in the long term. Hearing loss as a result of aminoglycoside dosage is delayed and may require an accumulation of aminoglycosides over a period of time [3]. A recent paper also suggests that the MET may act as a one-way valve for aminoglycosides, resulting in high concentration of cytosolic aminoglycosides [4]. Athough it is possible that higher concentrations of aminoglycosides may accumulate inside the cells which may disturb motility through secondary processes, therapeutic levels of aminoglycosides are expected to be well below the concentrations used in this study. This study suggests that aminoglycosides do not have any immediate or direct effect on OHC somatic motility.

Acknowledgment

This work was supported by NIH grant R01 DC004696 to D.H.

References

1. Guan, M., Fischel-Ghodsian, N., Attardi, G., 2000. A biochemical basis for the inherited susceptibility to aminoglycoside ototoxicity. Human Mol. Genetics 9, 1787-1793.
2. Duan, M., Agerman, K., Emfors, P., Canlon, B., 2000. Complementary roles of neurotrophin 3 and a *N*-methyl-D-aspartate antagonist in the protection of noise and aminoglycoside-induced ototoxicity. Proc. Natl. Acad. Sci. USA 97, 7597-7602.
3. Hashino, E., Shero, M., Salvi, R., 2000. Lysosomal augmentation during aminoglycoside uptake in cochlear hair cells. Brain Res. 887, 90-97.
4. Marcotti, W., van Netten, S.M., Kros, C.J., 2005. The aminoglycoside antibiotic dighyrodstreptomycin rapidly enters mouse outer hair cells through the mechano-electrical transducer channel. J. Physiol. 567, 505-521.
5. Takumida, M., Anniko, M., 2001. Nitric oxide in guinea pig vestibular sensory cells following gentamicin exposure in vitro. Acta Otolaryngol. 121, 346-350.
6. Lesniak, W., Pecoraro, V., Schacht, J., 2005. Ternary complexes of gentamicin with iron and lipid catalyze formation of reactive oxygen species. Chem. Res. Toxicol. 18, 357-364.
7. Jiang, H., Sha, S., Schacht, J., 2005. NF-κB pathway protects cochlear hair cells from aminoglycoside-induced ototoxicity. J. Neurosci. 79, 644-651.
8. Albinger-Hegyi, A., Hegyi, I., Nagy, I., Bodmer, M., Schmid, S., Bodmer, D., 2006. Alteration of activator protein 1 DNA binding activity in gentamicin-induced hair cell degeneration. Neurosci. 137, 971-980.
9. Silverblatt, F.J., Kuehn, C., 1979. Autoradiography of gentamicin uptake by the rat proximal tubule cell. Kidney Int. 15, 335-345.
10. Dulon, D., Hiel, H., Aurousseau, C., Erre, J.P., Aran, J.M., 1993. Pharmacokinetics of gentamicin in the sensory hair cells of the organ of Corti: rapid uptake and long term persistence. C. R. Acad. Sci. III. 316, 682-687.
11. Vital brazil, O., Prado-Franceschi, J., 1969. The neuromuscular blocking action of gentamicin. Arch. Int. Pharmacodyn. Ther. 179, 65-77.
12. Prado, W.A., Corrado, A.P., Marseillan, R.F., 1978. Competitive antagonism between calcium and antibiotics at the neuromuscular junction. Arch. Int. Pharmacodyn. Ther. 231, 297-307.
13. Ohmori, H., 1985, Mechano-electrical transduction currents in isolated vestibular hair cells of the chick. J. Physiol.359,189-217.
14. Kroese, A.B., Das, A., Hudspeth, A.J., 1989. Blockage of the transduction channels of hair cells in the bullfrog's sacculus by aminoglycoside antibiotics. Hear. Res. 37, 203-217.

15. Lin X., Hume R., Nuttal A., 1993. Voltage-dependent block by neomycin of the ATP-induced whole cell current of guinea-pig outer hair cells. J Neurophysiol. 70, 1593-1605.
16. Blanchet C., Erostegui, C., Sugasawa, M., Dulon, D., 2000. Gentamicin blocks ACh-evoked K+ current in guinea-pig outer hair cells by impairing Ca 2+ entry at the cholinergic receptor. J. Physiol. 525, 641-654.
17. Amici, M., Eusebi, F., Miledi, R., 2005. Effects of the antibiotic gentamicin on nicotinic acetylcholine receptors. Neuropharmacology 49, 627-637.
18. Brownell, W.E., Bader, D., Ribaupierre, Y., 1985. Evoked mechanical responses of isolated cochlear outer hair cells, Science 227, 194-196.
19. Kachar, B., Brownell, W.E., Altschuler, R., Fex, J., 1986. Electrokinetic shape changes of cochlear outer hair cells. Nature 322, 365-368.
20. He, D.Z.Z., Dallos, P., 1999. Somatic stiffness of cochlear outer hair cells is voltage-dependent. Proc. Natl. Acad. Sci. USA 96, 8223-8228.
21. Liberman, M.C., Gao, J., He, D.Z.Z., Wu, X., Jia, S., Zuo, J., 2002. Prestin is required for electromotility of the outer hair cell and for the cochlear amplifier. Nature 419, 300-304.
22. Dallos, P., Evans, B.N., Hallworth, R., 1991. Nature of the motor element in electrokinetic shape changes of cochlear outer hair cells. Nature 350, 155-157.
23. Kalinec, F., Holley, M.C., Iwasa, K.H., Lim, D.J., Kachar, B., 1992. A membrane-based force generation mechanism in auditory sensory cells. Proc. Natl. Acad. Sci. USA 89, 8671-8675.
24. Huang, G., Santos-Sacchi, J., 1994. Motility voltage sensor of the outer hair cell resides within the lateral plasma membrane. Proc. Natl. Acad. Sci. USA 91,12268-12272.
25. Zheng, J., Shen, W., He, D.Z.Z., Long, K.B., Madison, L.D., Dallos, P., 2000. Prestin is the motor protein of cochlear outer hair cells, Nature 405, 149-155.
26. Oliver, D., He, D.Z.Z., Klöcker, N., Ludwig, J., Schulte, U., Waldegger, S., Ruppersberg, J.P., Dallos, P., Fakler, B., 2001. Intracellular anions as the voltage sensor of prestin, the outer hair cell motor protein. Science 292, 2340-2343.
27. Nakashima, T., Teranishi, M. Hibi, T., Kbayashi, M., Umemura, M., 2000. Vestibular and chochlear toxicity of aminoglycosides – a review. Acta Otolaryngol. 120, 904-911.
28. He, D.Z.Z., Evans, B.N., Dallos, P., 1994. First appearance and development of electromotility in neonatal gerbil outer hair cells. Hear. Res. 78, 77-90.
29. Santos-Sacchi, J., Kakehata, S., Takahashi, S., 1998. Effects of membrane potential on the voltage dependence of motility-related charge in outer hair cells of the guinea-pig. J. Physiol. 510, 225-235.
30. Ashmore, J., 1989. Transducer motor coupling in cochlear outer hair cells. Mechanics of Hearing. Kemp, D., Wilson, J.P., editors, pp 107-113, Plenum Press, New York.

31. Santos-Sacchi J., 1991. Reversible Inhibition of Voltage-dependent Outer Hair Cell Motility and Capacitance. J. Neurosci. 11, 3096-3110.
32. Armstrong, C.M., Bezanilla, F., 1974. Charge movement associated with the opening and closing of the activation gates of the Na channels. J. Gen. Physiol. 63, 533-552.
33. Kakehata, S., Santos-Sacchi, J., 1996. Effects of salicylate and lanthanides on outer hair cell motility and associated gating charge. J. Neurosci. 16, 4881-4889.
34. He, D.Z.Z., Jia, S.P., Dallos, P., 2003. Prestin and the dynamic stiffness of cochlear outer hair cells. J. Neurosci. 23, 9089-9096.
35. Nomura, K., Naruse, K., Watanabe, K., Sokabe, M., 1990. Aminoglycoside blockade of Ca2(+)-activated K+ channel from rat brain synaptosomal membranes incorporated into planar bilayers. J. Membr. Biol. 115, 241-251.
36. Pichler, M., Wang, Z., Grabner-Weiss, C., Reimer, D., Hering, S., Grabner, M., Glossmann, H., Striessnig, J., 1996. Block of P/Q-type calcium channels by therapeutic concentrations of aminoglycoside antibiotics. Biochemistry 35, 14659-14664.
37. Mead, F., Williams, A., 2004. Electrostatic mechanisms underlie neomycin block of the cardiac ryanodine receptor channel (RyR2). Biophys. J. 87, 3814-3825.
38. Brasseur, R. Laurent, G., Ruysschaert, J.M., Tulkens, P., 1984. Interactions of aminoglycoside antibiotics with negatively charged lipid layers. Biochemical and conformational studies. Biochem. Pharmacol. 33, 629-637.
39. William, S.E., Zenner, H.P., Schacht, J., 1987. Three molecular steps of aminoglycoside ototoxicity demonstrated in outer hair cells. Hear. Res. 30, 11-18.

MECHANOTRANSDUCTION IN BONE CELL NETWORKS

X. E. GUO AND E. TAKAI

Department of Biomdical Engineering, Columbia University,
1210 Amsterdam Ave, New York, NY 10027, USA
E-mail: exg1@columbia.edu

X. JIANG, Q. XU AND G. M. WHITESIDES

Department of Chemistry and Biological Chemistry, Harvard University,
12 Oxford Street, Cambridge, MA 02138, USA

J. T. YARDLEY

Nanoscale Science and Engineering Center, Columbia University,
530 West 120th Street, New York, NY 10027, USA

C. T. HUNG

Department of Biomdical Engineering, Columbia University,
1210 Amsterdam Ave, New York, NY 10027, USA

K. D. COSTA

Department of Biomdical Engineering, Columbia University,
1210 Amsterdam Ave, New York, NY 10027, USA

Although osteocytes, the mechanosensor cells of bone tissue, form well organized interconnected cellular networks, most *in vitro* studies of bone cell mechanotransduction use uncontrolled monolayer cultures. In this study, bone cells were successfully cultured into a micropatterned network with dimensions close to that of in vivo osteocyte networks using microcontact printing and self-assembled monolyers (SAMs). The optimal geometric parameters for the formation of these networks were determined in terms of circle diameters and line widths. Bone cells patterned in these networks were also able to form gap junctions with each other, shown by immunofluorescence staining for the gap junction protein connexin 43, as well as the transfer of gap-junction permeable calcein-AM dye. We have demonstrated for the first time, that the intracellular calcium response of a single bone cell indented in this bone cell network, can be transmitted to neighboring bone cells through multiple calcium waves. Furthermore, the propagation of these calcium waves was diminished with increased cell separation distance. Thus, this study provides new experimental data that support the idea of osteocyte network memory of mechanical loading similar to memory in neural networks.

(A major portion of this chapter has been published in Molecular and Cellular Biomechanics, vol. 3(3):95-107, 2006)

1 Introduction

Osteocytes are interconnected through numerous intercellular processes, forming extensive cell networks throughout the bone tissue [1, 2]. Although it has been shown that osteocyte density is an important physiological parameter, with a decrease in osteocyte density with age and microdamage accumulation [3], most studies on osteocyte mechanotransduction have been performed on confluent or sub-confluent uncontrolled monolayers of bone cells [4-7]. Also, a decrease in osteocyte connectivity and disruption of their spatial distribution has been observed in osteoporotic bone [8]. Therefore, the ability to culture bone cells in a controlled network configuration, by modification of the surface chemistry, with prescribed cell separation distances and/or connectivity, would give insight into the response of bone cell networks to mechanical stimulation, in a scale that is more physiologically relevant than previously possible. Controlled bone cell culturing using micropatterning techniques, in combination with atomic force microscopy, which allows targetd stimulation of single cells within the network, would also provide a venue to study signal propagation between a single stimulated bone cell to neighboring bone cells in this controlled cell network.

In vivo, osteocyte bodies reside in lacunae approximately 10-15 μm in diameter, and connect to a maximum of 12 neighboring osteocytes through smaller channels (canaliculi) 0.2-0.8 μm in diameter and 15-50 μm long [9-11], in a 3-dimensional network. In vitro, bone cells generally exhibit high adhesion to many surfaces, and over time they can secrete their own extracellular matrix (ECM) proteins to modify the characteristics of the surfaces to which they adhere [12]. Therefore, in order to micropattern bone cells in a 2-dimensional network with feature sizes that are close to that of canaliculi (<1 μm) and lacunae (~20 μm), well-controlled surface chemistry is necessary. Previously, bovine and human endothelial cells, hepatocytes, and fibroblasts have been successfully micropatterned using self-assembled monolayers (SAMs) and soft lithography techniques (microcontact printing), into lines as thin as 10 μm wide, and islands as small as 10 μm x 10 μm [13-17]. SAMs spontaneously form ordered aggregates on metal-coated surfaces (*e.g.*, gold, platinum), and SAM modified surfaces allow strict control of cell-surface interactions through the creation of micropatterns of ECM proteins, surrounded by non-adhesive SAM regions such that individual cells will attach and spread only to the ECM patterned adhesive regions. The micropatterning of SAMs can be accomplished either by microcontact printing using polydimethyl siloxane (PDMS) elastomeric stamps created using soft lithography [15-17], or by gold lift-off techniques [18, 19]. In both techniques, alkanethiol SAMs can be micropatterned on gold coated surfaces, which have previously been used to control the interactions of surfaces with proteins [16, 20]. Hydrophobic alkanethiol SAMs such as octadecanethiol ($HS\text{-}(CH_2)_{17}CH_3$) rapidly and irreversibly adsorb proteins and promote cell adhesion, while SAMs that present ethylene glycol moieties such as tris-(ethylene glycol)-terminated alkanethiols ($HS-(CH_2)_{11}(OCH_2CH_2)_3OH$) effectively resist protein absorption and cell adhesion [15, 17, 18, 20-23]. Thus, the patterning of these two SAMs on a substrate defines the pattern of ECM proteins that are adsorbed from solution onto the substrate, and a grid of adhesive ECM islands and lines limits cell attachment to those islands. Although other modified silanes have been used to pattern bone cells [18, 24] into relatively thick lines or islands (several cells wide), alkanethiol SAMs have not been previously used to pattern bone cells, and bone cells have never been cultured in network patterns that closely mimic osteocyte networks *in vivo*.

The goals of this study were to 1) optimize the geometric parameters to create bone cell networks, 2) examine calcium wave propagation from a single bone cell indented using an atomic force microscope (AFM) to neighboring cells in this bone cell network, and 3) examine the effect of separation distance on calcium signal propagation.

2 Materials and Methods

2.1 Microcontact printing for the formation of controlled bone cell networks

Fibronectin (FN) patterns were created on gold-coated coverslips using microcontact printing techniques with SAMs and a PDMS elastomeric stamp, in a similar manner to Chen et al., [15]. Briefly, a mold was fabricated using Shipley 1818 positive photoresist (MicroChem Corp, Newton, MA) by spin-coating a 2 μm thick film of photoresist onto silicon wafers and exposing the photoresist to UV light through a chromium mask containing the desired grid and circle features (Fig. 1). The photoresist was then developed in a commercial Shipley photoresist developer, and exposed to a vapor of (tridecafluoro 1,1,2,2 tetrahydro octyl)-1-trichlorosilane to facilitate easy removal of the PDMS from the master. A 10:1 mixture of PDMS pre-polymer and curing agent (Sylgard 184 kit, Dow Corning, Midland, MI) was then prepared, poured onto the master, and placed under a vacuum to evacuate all air bubbles. The PDMS mixture was then cured at 70ºC for 2-4 hours and removed from the master such that the PDMS stamp contained the raised circle and grid micropatterns (Fig. 1 inset). To initially determine the optimal geometric parameters for network pattern formation, a mask with line widths varied as 1, 2, or 3 μm and the circle diameters ranged as 10, 15, 20, or 25 μm, with a fixed cell separation distance of 50 μm was used. Then to examine the effects of separation distance on signal propagation, another mask containing lines of 2 μm width and circles of 15 or 20 μm in diameter, with varying separation distances of 25, 50, or 75 μm was used.

Coverslips 48x65 mm were coated with a 10-15 Å adhesion layer of titanium and ~150 Å of gold using an electron-beam evaporator (Semicore SC200;

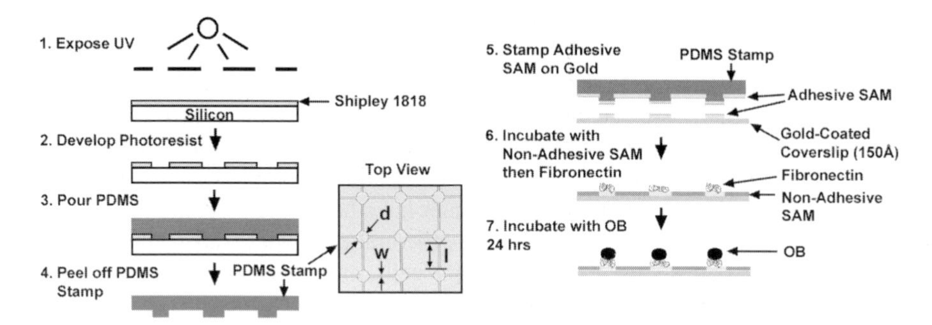

Figure 1. Flowchart of PDMS stamp fabrication, microcontact printing, and cell patterning. Insert is a top view photomicrograph of the PDMS stamp. Line width (w) of the patterns varies as 1, 2, or 3 μm, and circle diameter (d) varies as 10, 15, 20, or 25 μm, with a fixed separation distance (l) of 50 μm. Another similar stamp contained line widths fixed at 2 μm and circles of 15 or 20 μm diameter, with varied separation distances as 25, 50, or 75 μm.

Livermore, CA). The PDMS stamp was then coated with octadecanethiol (adhesive SAM; Sigma-Aldrich Co., St. Louis, MO), which allows cell adhesion, dried for 30 seconds under a gentle stream of nitrogen, and pressed onto the gold coverslips for 60 seconds (Fig. 1). The stamped coverslips were then immersed in an ethylene glycol terminated SAM solution (HS-C11-EG3, non-adhesive SAM; Prochimia, Sopot, Poland) for 1-3 hours to prevent cell adhesion to areas that were not patterned with the adhesive SAM. The patterned coverslips were then rinsed, dried under nitrogen, and further incubated with a 10 μg/ml solution of fibronectin (FN, Invitrogen, Carlsbad, CA), which was only absorbed to the adhesive SAM patterned regions. Osteoblast-like MC3T3-E1 cells were then seeded on these patterned coverslips at a density of 1.0x104 cell/cm^2 and cultured in α-minimum essential medium (α–MEM) supplemented with 2% charcoal-stripped fetal bovine serum (CS-FBS; Hyclone Laboratories Inc., Logan, UT) and allowed to migrate onto patterns for 24 hours.

2.2 Optimization of geometric parameters for bone cell network formation

To confirm good micropatterning, coverslips patterned with FN but not seeded with cells were subjected to immunofluorescence staining for FN using an anti-FN primary antibody (Chemicon, Temecula, CA) and a FITC conjugated secondary antibody (ICN/Cappel, Aurora, OH). Images of the stained coverslips were obtained using an inverted fluorescence microscope (Olympus IX-70, Melville, NY) with a 40x objective. Also, to confirm proper fabrication of the stamp features, the nominal features of stamps with a fixed 50 μm separation and variable line widths and circle diameters were measured 3 times and averaged from images obtained using a light microscope and Scion Image (Frederick, MD), image analysis software.

To determine the most optimal line widths and circle diameter sizes for the formation of bone cell networks, MC3T3-E1 cells were patterned as described above using stamps with a fixed separation distance of 50 μm, then fixed in 10% buffered formalin, and subjected to immunofluorescence staining against fibronectin and counterstained with a propidium iodide nucleic acid counterstain (Molecular Probes, Eugene, OR). An area 15x15 cells in the center of the patterned regions was analyzed. The total number of cells in the correct place (nucleus/cell body in the circle regions) and incorrect place were manually counted to give the fraction of cells in the correct place/cells in the incorrect place. To assess connectivity, the number of nodes and branches were counted manually, and connectivity was defined to be $\chi = \varepsilon$/nodes, where ε is the Euler number which is defined as ε = number of nodes – number of branches [25]. The Euler number was divided by the number of nodes in the analyzed area in order to obtain a connectivity measure that is independent of sample area size. The ideal connectivity for a network of 4 adjacent neighbors is -1.

2.3 Assessment of gap junction formation

Immunofluorescence staining against connexin 43 (Cx43) was performed on bone cells patterned as described above, which were then fixed with cold acetone for 20 minutes at -20°C. To visualize gap junctions, the osteoblasts on the patterned glass coverslips were incubated with a polyclonal anti-Cx43 (Chemicon) primary antibody followed with a FITC-conjugated anti-rabbit (Molecular Probes) secondary antibody, then counterstained with propidium iodide nucleic acid counterstain. Samples were then examined using a fluorescence microscope with a 60x objective lens.

To assess the formation of functional gap junctions, a technique employing calcein dye transfer from fluorescently double labeled cells was used [7, 26]. Bone cells were trypsinized and stained with 5 μM 1,1'-dioctadecyl-3,3,3',3'-tetramethylindocarbocyanine perchlorate (DiI, Molecular Probes) for 20 minutes, followed by 4 μM calcein acetoxymethyl ester (calcein-AM, Molecular Probes) for 30 minutes. DiI is a membrane-bound dye that will not transfer to neighboring cells, thus serving as an indicator of the original double labeled cells, while calcein is a gap junction permeable dye. The double labeled cells were then mixed with a suspension of unlabeled bone cells at a ratio of 1:80 and cultured on the fibronectin patterned coverslips as described above. After culturing the cells overnight, the patterned cells were imaged using a fluorescence microscope equipped with a rhodamine filter (red) to visualize the original double labeled cells, and a fluorescein filter (green) to visualize neighboring cells that received the calcein dye through gap junctional coupling to the double labeled cells.

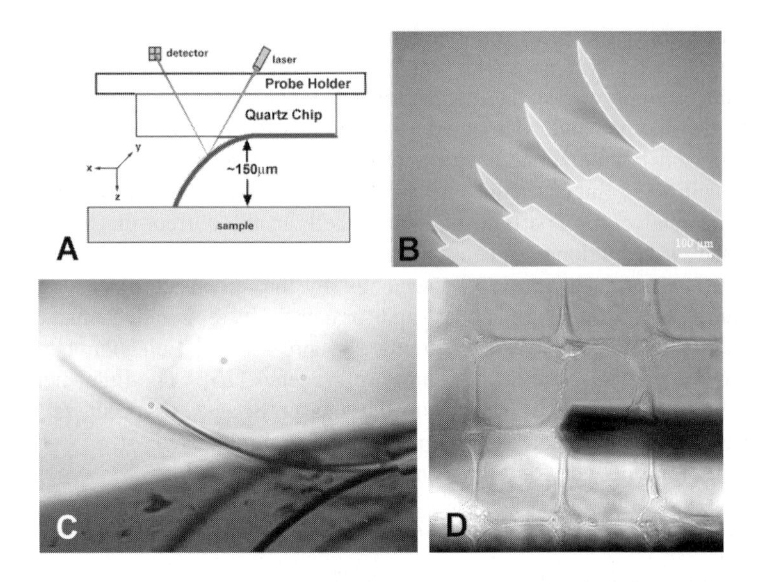

Figure 2. StressedMetal™ probe. A) Side-view illustration of the longest probe indenting a sample; B) SEM image of probes; C) Side view light micrograph of longest and second longest probes; D) Top view light micrograph of the longest probe indenting a single bone cell in the network. A and B are adopted from www.parc.com.

2.4 Single-cell nanoindentation using atomic force microscopy

To examine the effects of cell separation distance on calcium wave propagation from a single indented bone cell, bone cells were cultured into patterns with 20 μm or 15 μm diameter circles and 2 μm wide lines, with a separation distance of 50, or 75 μm, according to the procedures described above. Patterned cells were loaded with Fluo-4 AM (Molecular Probes, Eugene, OR), a fluorescence calcium indicator dye, by incubating the cells in a solution containing 5 μM Fluo-4 AM, 0.02% pluronic F-127 for even dispersion of the dye, and α-MEM supplemented with 0.5% CS-FBS, for 2 hours at room temperature. The glass coverslips containing the patterned bone cells were then placed under an AFM (Bioscope, Digital Instruments/Veeco, Santa Barbara, CA) mounted on an inverted fluorescence microscope (Olympus IX-70) equipped with a cooled digital CCD camera (Sensicam, Cooke Corp, Auburn Hills, MI), and allowed to equilibrate to ambient conditions for ~5 minutes. The AFM was mounted with a specialized probe with an extremely high aspect ratio (StressedMetal™, Palo Alto Research Center, Palo Alto, CA), ~150 μm high x 400 μm long, to minimize fluid motion at the cell surface due to the probe holder displacement (Fig. 2) [27, 28]. The stiffness of the probe was 0.06 N/m [29]. These stress-engineered cantilever probes are unique because they are much taller than conventional probes and fabricated on optically transparent substrates, which enables straightforward measurements inside fluids because the fluid is flush against the glass. In addition, the stress-engineered cantilever tips can be hundreds of microns away from the AFM probe holder and parallel to the sample substrate. In contrast conventional cantilevers are only tens of microns tall. A custom probe holder was made such that the StressedMetal probe was held at a 0° angle with the contact surface, rather than the 12° angle of conventional probe holders, to optimize the reflection of the laser spot on the probe. A single bone cell in the network pattern was stimulated by continuous indentation at 1 Hz, in relative trigger mode, to apply a prescribed contact force of 61.6±13 nN, which resulted in an 823±380 nm indentation depth (locally). The AFM indentation experiment involves monitoring the deflection of the StressedMetal probe as its tip contacts and indents the cell. The resulting interaction force bends the probe, which is detected by the movement of a reflected laser spot.

Simultaneously with the indentation, fluorescence time-lapsed images of intracellular calcium ($[Ca^{2+}]_i$) waves in bone cells were taken every 2 seconds, starting 60 seconds prior to stimulation, to obtain baseline $[Ca^{2+}]_i$ levels, to up to 5 minutes after the start of the stimulation. Images were analyzed using MetaMorph 4.1™ imaging software (Universal Imaging Corp., West Chester PA), where the mean fluorescence intensities of individual cells were measured and background fluorescence was subtracted for each time-lapsed image. The relative change in intracellular calcium was determined by dividing the fluorescence measurement of each cell after stimulation by the average baseline fluorescence intensities of each cell prior to stimulation. The response of individual bone cells in a field of view, using a 20x objective, was analyzed, thereby permitting the analysis of individual cell responses. A responsive cell was defined as a cell with a calcium oscillation of at least four times the maximum oscillation measured during the baseline measurement period immediately before stimulation [30]. The speed of calcium wave transmission was also determined by dividing the distance between the indented cell and neighboring responding cells by the time between the response of the indented cell and neighboring responding cells. The percentage of responding cells immediately adjacent and two cells away, were also determined. Five experiments were performed for each separation distance, with a total of 30 and 24 responding cells analyzed in cell networks with 50 and 75 µm separation distances, respectively. The calcium signal transmission speed from the indented bone cell to neighboring bone cells of various cell steps (e.g. one cell away, two cells away) and differences in the percentage of responding cells with each cell step at different separation distances was compared. The peak calcium response magnitudes of cells, between different separation distances at each cell step were also analyzed. To determine statistical significance between different conditions a two-way ANOVA with a Fisher's post-hoc analysis (Systat, Point Richmond, CA) was used. For all statistical analyses a p value of less than 0.05 was considered significant.

3 Results

3.1 Assessment of cell patterning

Measurements of the PDMS stamp features using Scion Image showed that the nominal dimensions of the fabricated stamps were slightly smaller than the prescribed feature dimensions by a maximum of 18.1% for line widths and 2.0% for circle diameters (Table 1). Variations in the nominal dimensions were due to slight alterations in photoresist thickness, PDMS shrinkage, or UV exposure time. Fluorescent micrographs of patterned coverslips stained for fibronectin showed consistent good pattern transfer for all feature dimensions (Fig. 3). Transferred pattern dimensions were larger than the nominal stamp dimensions by a maximum 27.2% for line widths and 28.8% for circle diameters.

Qualitatively, good pattern formation was achieved, where the majority of cell bodies reside in the circles and the cell processes extend along the lines (Fig. 3). Quantitative assessment of pattern formation revealed that, in general, features with 2 μm wide lines had the highest fraction of cells in correct/incorrect locations, while w=2 μm x d=20 μm and w=3 μm x d=15 μm showed the best connectivity (Table 2). In contrast, larger circle diameters and larger line widths (w=3 μm x d=20-25 μm) led to least optimal patterning with many cells adhering to the line areas, and significantly lower correct/incorrect cell positioning ratios. For patterns with varying separation distances, bone cells could only be successfully cultured into network patterns at 50 and 75 μm separation distances but not 25 μm separation distance, since this shorter distance allowed cell bodies to span over multiple circles, thus preventing good network pattern formation (Fig. 4).

Immunofluorescence staining for Cx43 gap junction protein showed punctate Cx43 staining at the ends of cell processes, which suggests that gap junctions were formed between cells in the micropatterned bone cell network (Fig. 4D). The calcein dye transfer assay showed that neighboring cells at least 1 to 2 cell steps away from the original double labeled cells were able to receive the calcein dye, demonstrating that functional gap junctions form between bone cells in the network patterns (Fig. 5). The calcein dye transfer was similar in bone cell networks with 50 and 75 μm separation distances.

Table 1. Nominal dimensions of the PDMS stamps and percent error of nominal dimensions from the prescribed dimensions.

	Prescribed Dimension	Nominal Dimension	Standard Deviation	Percent Error
Line Width (μm)	1	0.99	0.1	−0.7
	2	1.80	0.2	−10.0
	3	2.46	0.3	−18.1
Circle Diameter (μm)	10	10.17	0.3	+1.7
	15	14.79	0.2	−1.4
	20	19.59	0.3	−2.0
	25	24.50	0.4	−2.0

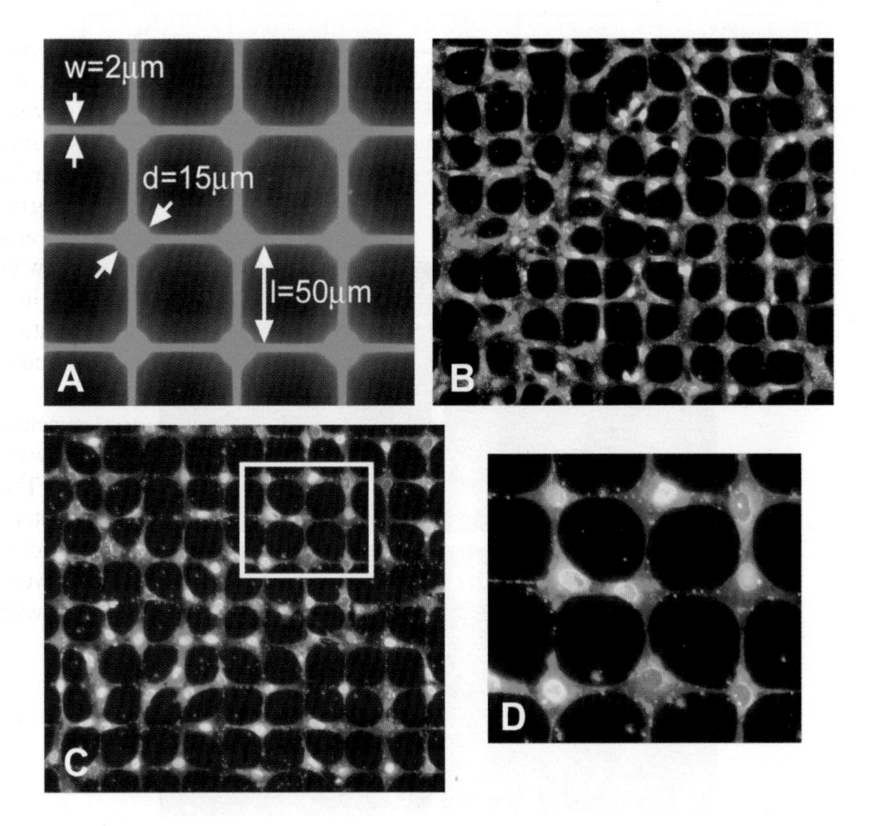

Figure 3. A) Fluorescent micrograph of a typical micropatterned coverslip stained against fibronectin. Micropatterned features of other dimensions were similar. B-D) Micropatterned bone cells stained with fibronectin (green) and propidium iodide nuclear counterstain (red). B) Least optimal pattern with w=3 μm x d=25 μm. C) Optimal pattern w=2 μm x d=20 μm; D) enlargement of C) illustrating that cell bodies reside in circles and processes extend along lines.

Table 2. Quantitative assessment of bone cell pattern formation with varying feature dimensions. The ideal connectivity for a network of 4 adjacent neighbors is −1.

Dimensions	%Correct of Total Cells	Correct/Incorrect	Euler #	Connectivity
1μmx10μm	76.0	3.5	−193.3	−1.34
1μmx15μm	78.9	3.8	−184.2	−1.47
1μmx20μm	73.1	3.0	−233.7	−1.46
1μmx25μm	72.2	3.2	−242.3	−1.38
2μmx10μm	82.5	4.7	−212.7	−1.40
2μmx15μm	83.0	4.9	−190.3	−1.32
2μmx20μm	81.6	4.5	−195.2	−1.20
2μmx25μm	77.4	3.5	−205.8	−1.28
3μmx10μm	74.8	3.2	−226.2	−1.24
3μmx15μm	81.2	4.9	−200.8	−1.18
3μmx20μm	58.4	1.4	−326.2	−1.56
3μmx25μm	68.1	2.8	−239.8	−1.34

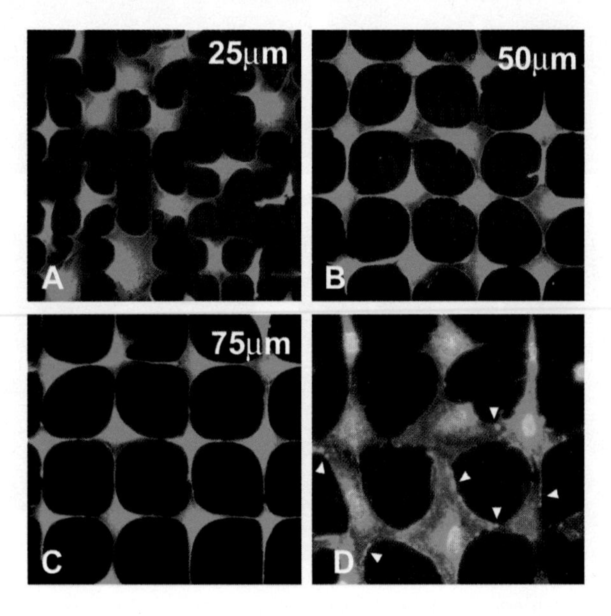

Figure 4. A-C) Micropatterned bone cells with 25, 50, or 75 μm intercellular separation distances stained with eosin (green). 50 and 75 μm separation distances support good pattern formation, while a 25 μm separation allows bone cells to span multiple spots. D) Connexin 43 (Cx43) staining (green) and propidium iodide nucleic acid counterstain (red). Arrowheads show areas of high punctate Cx43 staining between adjacent cells, suggesting gap junction formation.

3.2 Calcium wave propagation in bone cell networks

Intracellular calcium signals were observed to propagate from a single stimulated bone cell (cell #1) to adjacent cells micropatterned in the network configuration at both 50 and 75μm separation distances (Fig. 6). Some neighboring cells (#3 and #5) were able to respond with a second calcium transient through different cell paths. For example, cell #3 first showed a response propagated through cell #2, then a second response through cells # 6 and #7. Similarly, cell #5 responded first through the indented cell (#1), then responded for a second time through cells #6 and #9. The magnitudes of the second calcium transient of cells #3 and #5 were similar to the magnitudes of the first responses (Fig. 7), and in general, second responses were not smaller than the first responses in networks with 50 μm separation distances (Fig. 8). The second responses in networks with a 75 μm separation were smaller than the first responses. An average of 73.3±5 and 105.0±51 seconds elapsed between first and second responses for 50 μm (n=3) and 75 μm (n=2) separation distances, respectively.

There was no significant difference in transmission speed with increased cell steps (1 vs. 2 cells away) or with increased separation distance (50 vs. 75 μm)

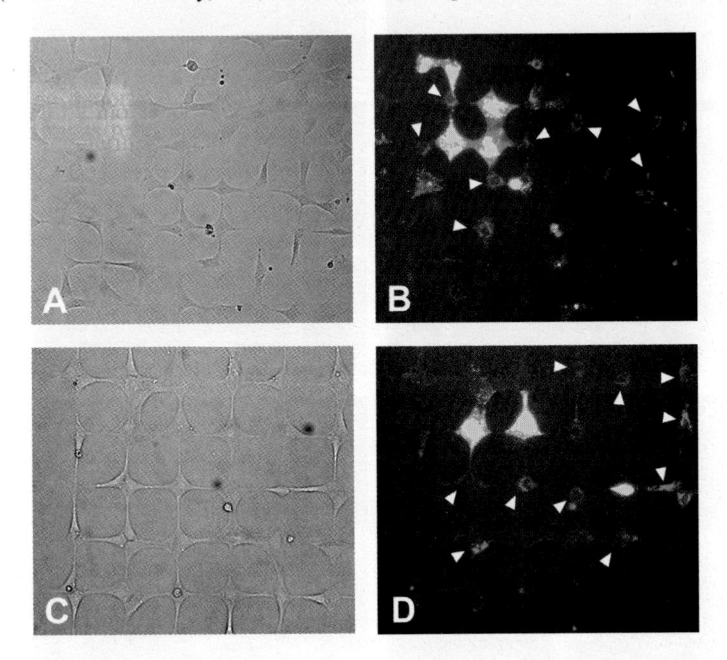

Figure 5. Calcein dye transfer assay to assess functional gap junction formation. Panels A and C are light micrographs, and B and D are fluorescent micrographs. A, B) 50 μm separation distance. C, D) 75 μm separation distance. Calcein dye (green, arrowheads) was transferred to neighboring cells at least 2 cell steps away from the original double labeled cells (red/yellow) in both bone cell networks of 50 and 75 μm separation distance.

(Fig. 9). The mean transmission speeds were 3.7±2.8 and 3.3±2.3 μm/sec for 50 and 75 μm separation distances, respectively. In addition, the magnitudes of the calcium responses of neighboring cells were significantly smaller than that of the indented cell (Fig. 9). However, there was no significant difference between the response magnitudes of cells 1 step or 2 steps away for either separation distance. There was a significantly smaller percentage of responsive cells 2 cell steps away in networks with a 75 μm separation distance compared to those with a 50 μm separation distance (Fig. 10). However, there was no difference in the percentage of responsive cells 1 cell step away, or directly adjacent to the indented bone cell, regardless of the cell separation distance.

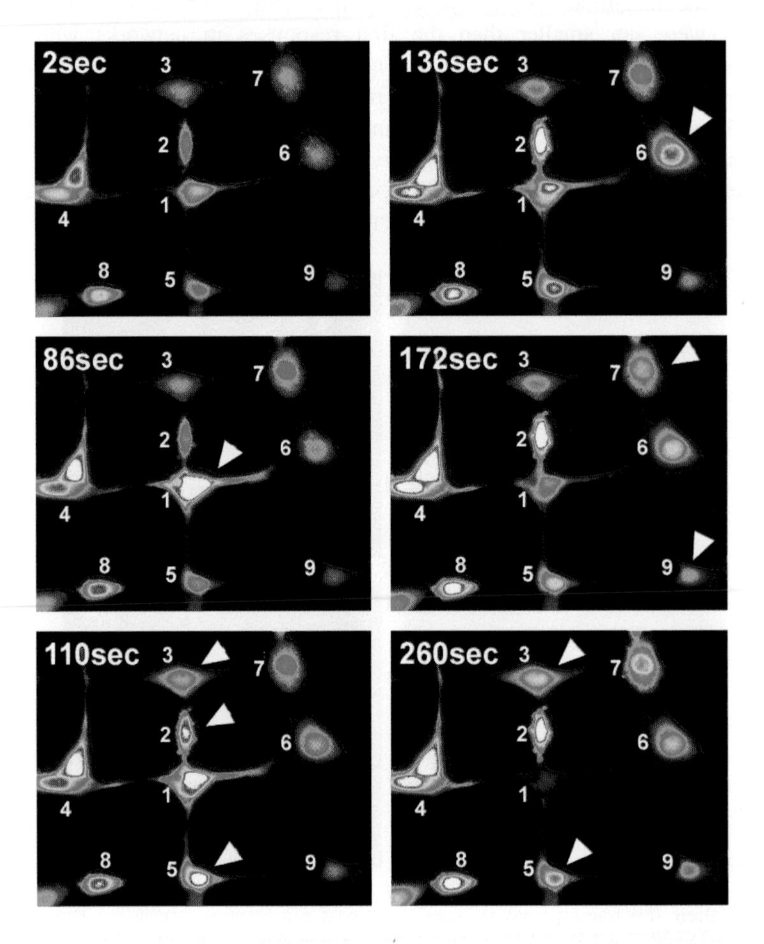

Figure 6. Calcium signal propagation from a single indented bone cell (#1) to adjacent cells in the network pattern with a 50 μm separation distance. Arrowheads highlight some responding cells. Cells #3 and #5 were able to respond twice through different cell pathways.

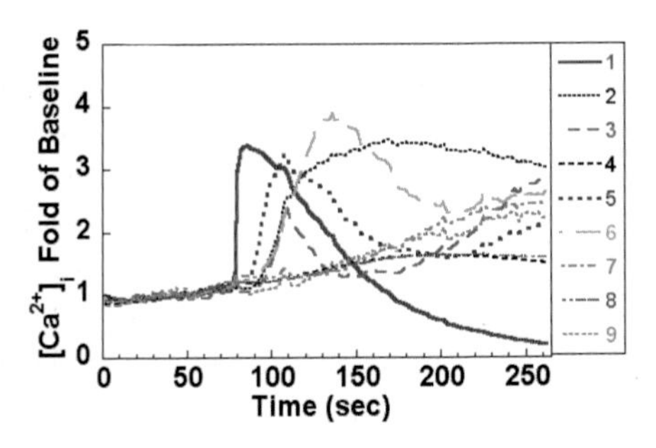

Figure 7. Calcium response expressed as the fold change in [Ca2+]i over baseline over time, corresponding to the cells in Figure 6. Cell #1 is the indented cell.

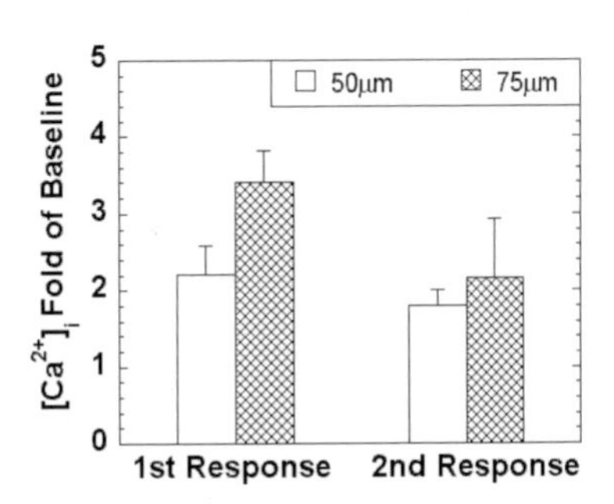

Figure 8. Magnitude of 1st and 2nd calcium responses expressed as a fold increase over baseline calcium measurements, for bone cells networks with 50 and 75 μm separation distances. n=3 for 50 μm and n=2 for 75 μm separation distance networks. Results are expressed as means ± standard deviations.

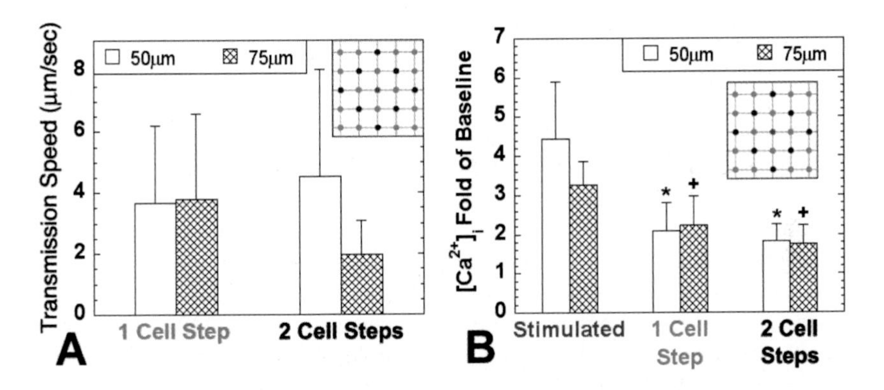

Figure 9. A) Transmission speed of calcium signal from the indented bone cell to bone cells 1 cell step and 2 cell steps away, in bone cell networks with separation distances of 50 μm or 75 μm. Inset: red = indented cell, green = 1 cell step away, black = 2 cell steps away. There is no significant change in transmission speed regardless of the number of cell steps or separation distances. B) Magnitude of the peak calcium response with each progressive cell step expressed as a fold increase over baseline calcium measurements. There is a significant decrease in calcium response between the indented cell and cells 1 or 2 steps away. *$p<0.001$ with stimulated cell in 50 μm separation distance network; +$p\leq0.01$ with stimulated cell in 75 μm separation distance network. There is no significant difference between the response magnitudes of cells 1 step or 2 steps away for either separation distance. Results are expressed as means ± standard deviations.

4 Discussion

In this study, bone cells were successfully cultured into a micropatterned network with dimensions close to that of in vivo osteocyte networks. The optimal geometric parameters for the formation of these networks were determined in terms of circle diameters and line widths. Bone cells patterned in these networks were also able to form gap junctions with each other, shown by immunofluorescence staining for the gap junction protein connexin 43, and transfer of the gap junction permeable calcein dye. Furthermore, we have demonstrated for the first time, that the intracellular calcium response of a single bone cell indented in this bone cell network, can be transmitted to neighboring bone cells through multiple calcium waves.

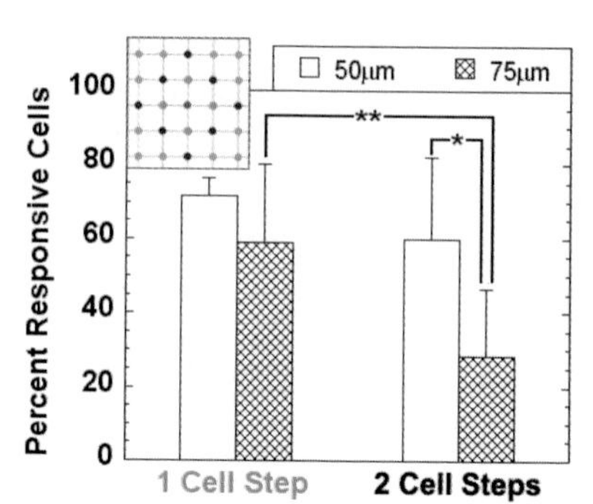

Figure 10. Percentage of responsive bone cells 1 cell step and 2 cell steps away, in bone cell networks with separation distances of 50 μm or 75 μm. Inset: red = indented cell, green = 1 cell step away, black = 2 cell steps away. There is a significantly smaller percentage of responsive cells 2 cell steps away in networks with a 75 μm separation distance compared to those with a 50 μm separation; *p=0.02. There was also a significant decrease in the percentage of responsive cells between 1 cell step and 2 cell steps away, in bone cell networks with a 75 μm separation distance; **p=0.01. Results are expressed as means ± standard deviations. n=5 experiments.

The formation of neural network circuits in the brain is the key to permanent memory in cognitive functions [31]. Osteocytes in mineralized bone tissue also form elaborate cellular networks. It is well known that mechanical usage modulates the shape, mass, and microstructure of bone. Does the osteocyte network hold the key to cellular memory of mechanical loading history in bone tissue? This is an interesting hypothesis which may have a profound implication in cellular and molecular mechanisms of bone adaptation to mechanical loading [32]. The current study (with osteoblast-like cells) may be suggestive of the potential for osteocyte network memory of mechanical loading reminiscent of neural networks.

The PDMS stamps were successfully fabricated with actual dimensions similar to the prescribed feature specifications. The dimensions of the patterned fibronectin were up to 29% larger than the stamp dimensions, and this enlargement may in part be due to lateral expansion of the raised stamp features due to the weight of the stamp. There may also be some systematic overestimation of patterned feature measurements due the difficulty in determining the edges of the fibronectin patterned features, which had a faint halo of fluorescence along the edges. Thus the actual pattern feature dimensions that promote optimal pattern formation are slightly larger than 15 or 20 μm circles x 2 μm wide lines.

Intracellular calcium transients were observed to be propagated from an indented bone cell to neighboring bone cells, both in networks with 50 and 75 μm separation distances. Some bone cells were able to exhibit double responses in intracellular calcium through signal propagation from two different cell paths. The time delay between the first and second responses ranged from approximately 65 to 150 seconds apart, similar to a previous finding that approximately 60-600 seconds elapsed between consecutive calcium responses of bone cells subjected to constant oscillatory fluid shear [30]. The time between responses were similar for both bone cells patterned with 50 and 75 μm separation distances. However, bone cells responding multiple times to fluid shear were shown to have decreased magnitude of subsequent responses compared to the initial response, while in the current study, second responses of bone cells were similar in magnitude to that of the first response. It is important to note that in the fluid shear study performed by Donahue *et al.*, all of the bone cells were stimulated with fluid shear, while in the current study the cells exhibiting second responses were not directly mechanically stimulated. Therefore, it is possible that the behavior of the calcium response differs between direct cell response to mechanical stimulation and propagated responses. Out of 10 experiments total, we were able to observe cells with multiple responses in 5 experiments (50%). The ability of bone cells to respond multiple times, without a decrease in the magnitude of the calcium response to transmitted calcium waves may play a role in memory of bone cell networks of their previous mechanical loading history. This mechanism may also have similarities with memory in neural networks [33, 34]. Thus, it would be of interest to examine whether there is an increase in gap junctional connections between bone cells along cell pathways with multiple responses in the network after mechanical stimulation. It is also possible that the multiple response behavior of bone cells in these networks modulate the signaling between osteocytic networks and osteoblasts on the surface of the bone.

The mean signal transmission speeds of 3.7±2.8 for 50 μm and 3.3±2.3 μm/sec for 75 μm separation distances measured in this study were similar to the signal transmission speed of ~2.5 μm/sec and ~0.5 μm/sec in osteoblasts (ROS 17/2.8) cultured in uncontrolled monolayers [35, 36]. Furthermore, the finding that signal propagation speed does not diminish with increased number of cell steps away from the indented bone cell, is in agreement with previous the findings of Xia and Ferrier [35], and suggests that the calcium signal may be regenerated to a certain extent at each cell. Also, even with increased cell separation distance from 50 to 75 μm, the signal transmission speed was not significantly reduced, supporting the idea that the signal transmission is not due to diffusion of secreted factors. Interestingly, a previous study of osteoblasts in nearly confluent monolayers showed that transmission of calcium signals through secretion of paracrine factors such as ATP is faster, with a transmission velocity of ~10 μm/sec, compared to gap junctional communication, with a transmission velocity of ~0.5 μm/sec [36].

The peak magnitudes of calcium responses in neighboring cells were significantly lower than that of the indented cell. However, the peak magnitudes were not decreased with transmission between other neighboring cells, such as cells 1 step away and 2 steps away. One possible explanation is that the mechanism of the calcium response between the indented cell and the neighboring non-indented cells may be different. Specifically, the calcium response of the indented cell may be through influx of external calcium, while the responses of non-indented cells may be dependent on the release of internal calcium stores. In support of this idea, previous studies using UMR 106-01 and HOBIT osteoblastic cells in uncontrolled monolayers have shown that depletion of internal calcium stores did not effect cell response to mechanical perturbation with a micropipette, but abolished the propagation of calcium response to neighboring cells [36, 37]. In contrast, removal of external calcium decreased the mechanical response of the indented cell but did not effect calcium signal propagation to neighboring cells [37].

Although the transmission speeds of calcium signals and the magnitudes of responses were not significantly different between bone cells in networks with 50 or 75 μm separation distances, the percentage of responsive cells 2 cell steps away was significantly lower in networks with 75 μm separation distance. It has been previously proposed that bone cell response to mechanical stimulation (fluid shear) of different magnitudes may be encoded by the percentage of responsive cells, where the percentage of responsive cells at any calcium responses magnitude threshold increases with the magnitude of stimulation [38]. Thus, signaling within osteocytic networks and to osteoblasts may also be dependent on the percentage of responsive cells. Reduced osteocyte density in aged, microdamaged, or osteoporotic bone [3, 8, 39, 40] increases separation distances between osteocytes, and may diminishes the percentage of responsive cells to mechanical loading of bone, thereby reducing the signaling to other bone cells on the surface. The reduction in bone cell signaling with increased cell separation distance or decreased cell density may explain the reduced mechanosensitivity of bone with age [41, 42].

For future studies, it would be necessary to repeat this study on osteocytes and to examine the interaction between the osteocyte network and osteoblasts. It would also be interesting to examine calcium propagation responses of bone cell networks subjected to different magnitudes of mechanical stimulation. Since the mechanism of the calcium wave propagation is not clear (e.g. secreted factors vs. gap junctional communication), studies to block gap junctional communication and/or paracrine signaling via ATP, nitric oxide, or prostaglandins would allow better characterization of the calcium signaling. Furthermore, exploration of the different mechanisms of calcium response, such as influx of external calcium or release of internal stores, in mechanically stimulated cells and those that received a transmitted calcium wave, would also be of interest. It would also be interesting to examine changes in calcium signal propagation with changes in cell connectivity, since osteoporotic bone has been shown to have decreased osteocyte connectivity. Since the activity of osteoblasts is known to be modulated by osteocytes [43], there may be modulation of calcium signal propagation between osteocytes and osteoblasts with alterations in osteocyte network connectivity and separation distances. Better understanding of signaling within osteocyte networks as well as between osteocyte networks and osteoblasts can provide information to devise treatments to enhance or manipulate bone adaptation.

Previous studies of mechano-signal transduction between osteoblasts and osteocytes have used intermixed monolayers of the two cell types, where the spatial distribution and organization of osteoblasts and osteocytes were uncontrolled [7]. Since osteocytes form networks with each other in the bone tissue and only interact with osteoblasts at the bone surface, only a small fraction of osteocytes in the bone tissue actually directly contact osteoblasts. Thus, in order to better understand mechanotransduction between osteoblasts and osteocytes, it is necessary to control the spatial orientation of osteoblasts and osteocytes. Manipulation of the spatial orientation and controlled interaction of different cell types can be achieved through modification of the surface chemistry using SAMs. In addition to the ability of ethylene glycol terminated non-adhesive SAMs to resist protein absorption, SAMs can be released from gold surfaces by passing a short cathodic voltage through the gold (electrochemical desorption). By releasing the non-adhesive SAMs, the surface chemistry of the areas that were previously unable to absorb proteins and sustain cell adhesion is altered, such that proteins can be absorbed and cells able to adhere to these regions [22]. Thus electrochemical desorption can be used to allow previously confined cell populations to migrate out of confinement and interact with other cells.

We have developed a two-dimensional micropattern system for co-culturing osteoblasts and osteocytes was created using gold lift-off techniques similar to those used by Healy *et al.* [18] and Sorribas *et al.* [19], and SAMs. Briefly, the micropattern was made on glass coverslips using a positive photoresist, Shipley 1818 (Fig. 11). The mask consisted of three 50x50 element grid patterns of circles and lines for the osteocytic network, and three 3.1 x 3.1 mm solid squares for osteoblasts. The patterned regions were separated by a 1 μm thick centerline and 1 μm thick vertical lines into six sub-regions (Fig. 11 top view), to electrically isolate the sub-regions for later electrochemical desorption of SAMs. Then ~150Å of gold was evaporated onto patterned coverslips using an e-beam evaporator onto the entire pattern. The Shipley photoresist was then removed by sonicating in ethanol, so that only the patterned regions were not covered by gold. A rectangular PDMS well, matching the border of the gold pattern, with a 100 μm divider was also created. This PDMS well was placed on the patterned cover slip such that the well is aligned with the border of the pattern, and the divider rests slightly above the centerline. Then, the wells were filled with an ethylene glycol terminated non-adhesive SAM, which assemble only in regions with gold, and incubated overnight. After the SAM solution was removed and rinsed with ethanol, the wells were then further incubated with 10 μg/ml fibronectin (FN), which is only absorbed to the bare glass regions. To create osteocyte-osteoblast co-cultures, the well containing the network patterns was seeded with an osteocyte-like MLO-Y4 cell suspension, and the well containing the solid squares was seeded with osteoblast-like MC3T3-

E1 cells (Fig. 12). In order to track the different cell types, 2 μM calcein-AM (Molecular Probes, Eugene, OR) was used to stain osteocytes, while osteoblasts were left unstained. After the cells were allowed to adhere for 24 hours non-adherent cells were gently washed away, and the divider was cut and removed. To release the osteoblasts from the confines of the square patterns, non-adhesive SAMs surrounding the square patterns were removed by passing a -1.8V current (180mV/mm) through the gold for 30 seconds [22]. Since lines of bare glass separated the gold patterned areas, the current could only travel within one sub-region so that the SAMs were only removed from that sub-region. Also, subjecting cells to a -1.8V current for 30 seconds has been found to have no effect on cell viability or motility [22]. Therefore, the osteocytes were still confined in their controlled network pattern, while the MC3T3-E1 cells were allowed to freely migrate toward the osteocytic network, and make connections with the network without crossing into it. Also, since the calcein dye in osteocytes was transferred through gap junctions, connections between osteoblasts and osteocytes could be observed (Fig. 12).

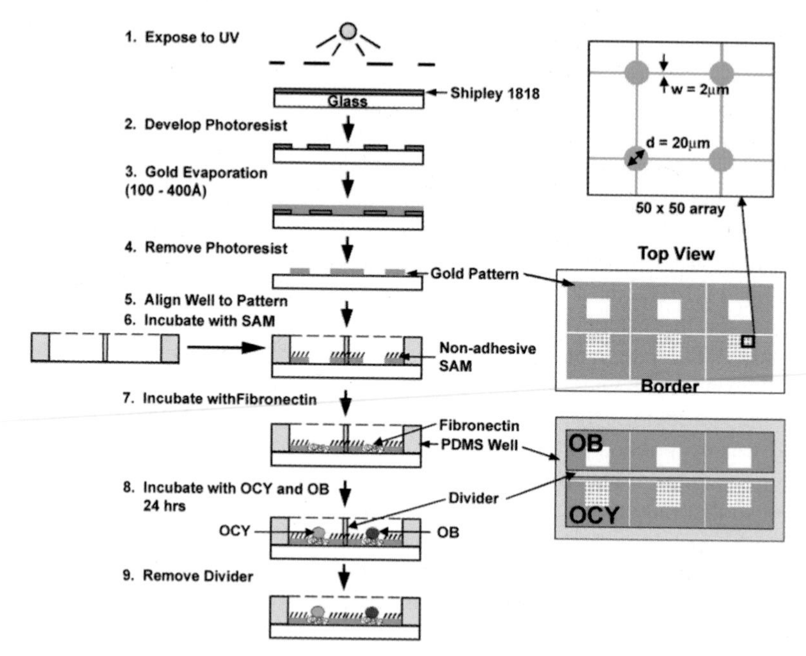

Figure 11. Flowchart of gold patterned coverslip fabrication using lift-off and co-culturing osteoblasts (OB) and osteocytes (OCY) on gold patterned coverslips using a PDMS well and SAM. Top view shows solid square regions for osteoblasts and grid patterned regions for osteocyte networks.

This controlled osteoblast-osteocyte co-culture can be used to study mechanotransduction between these two cell types. Specifically, the role of paracrine signaling in mechanotransduction between osteocytes and osteoblasts can be examined when the two cell types are confined to the square or micropatterned regions and cannot make direct contact. In addition signaling between the two cell types can be examined when they are in direct contact. Mechanotransduction with direct contact is likely to occur through gap junctional signaling, since osteocytes and osteoblasts have been shown to form gap junctions *in vivo* [44]. Controlled mechanical stimulation to just one cell population can be achieved through the use of an AFM. In addition, it may be possible to indent or compress one or more cells using a modified probe in conjunction with the AFM. Also, fluid shear stimulation of both cell populations, or the combination of fluid shear and chemical stimulation can be achieved through the use of multiple laminar flows through a microchannel [45]. Thus the controlled co-culture technique described here can provide a powerful tool to study signal transduction in osteocytes and osteoblasts in response to mechanical and chemical stimulation.

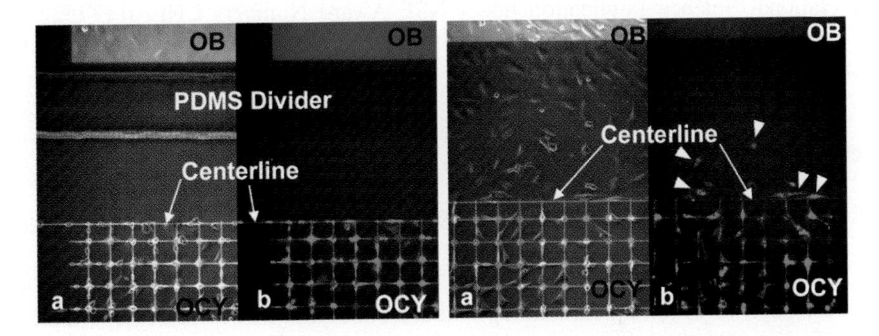

Figure 12. Light micrographs (a) and fluorescent micrographs (b) of osteoblasts and osteocytes in 2D co-culture. Co-culture before the release of osteoblasts (left) and co-culture after osteoblast release (right). Calcein dye transferred from osteocytes (OCY) to osteoblasts (OB) presumably via gap_junctions (arrowheads).

5 Conclusions

In this study, bone cells were cultured into micropatterned networks with controlled cell separation distances, and calcium wave propagation in this network in response to single cell indentation was examined. The conclusions are as follows:

1. Functional gap junctional coupling ocurrs between bone cells in micropatterned networks.
2. Calcium waves can be propagated from a single stimulated cell to neighboring cells, and these calcium waves can be propagated through different cell paths

without a decrease in transmitted calcium wave magnitudes, similar to neural networks.
3. Networks with increased cell separation distance, and thus decreased cell density, have a decreased percentage of responsive cells, suggesting a possible mechanism for decreased mechanosensitivity of bone with age.

Acknowledgments

The authors would like to thank Dr. Lynda Bonewald of the University of Missouri at Kansas City for her generous gift of MLO-Y4 cells. This work is support by NIH grants AR048287, AR049613 and AR052417 (XEG), AR046568 (CTH), and the NSF CAREER award BES00239138 (KDC). GMW, XJ and QX acknowledge NIH award GM065364. QX acknowledges support from NSF award PHY-0117795. This work is also supported by the Nanoscale Science and Engineering Initiative of the National Science Foundation under NSF Award Numbers CHE-0117752 and CHE-0641523, and by the New York State Office of Science, Technology, and Academic Research (NYSTAR).

References

1. Cowin, S.C., Weinbaum, S., 1998. Strain amplification in the bone mechanosensory system, Am. J. Med. Sci. 316, 184-188.
2. Weinbaum, S., Cowin, S.C., Zeng, Y., 1994. A model for the excitation of osteocytes by mechanical loading-induced bone fluid shear stresses, J. Biomech. 27, 339-360.
3. Vashishth, D., Verborgt, O., Divine, G., Schaffler, M.B., Fyhrie, D.P., 2000. Decline in osteocyte lacunar density in human cortical bone is associated with accumulation of microcracks with age, Bone 26, 375-380.
4. Cheng, B., Zhao, S., Luo, J., Sprague, E., Bonewald, L.F., Jiang, J.X., 2001. Expression of functional gap junctions and regulation by fluid flow in osteocyte-like MLO-Y4 cells, J. Bone Miner. Res. 16, 249-259.
5. Cheng, B., Kato, Y., Zhao, S., Luo, J., Sprague, E., Bonewald, L.F., Jiang, J.X., 2001. PGE(2) is essential for gap junction-mediated intercellular communication between osteocyte-like MLO-Y4 cells in response to mechanical strain, Endocrinology 142, 3464-3473.
6. Reilly, G.C., Haut, T.R., Yellowley, C.E., Donahue, H.J., Jacobs, C.R., 2003. Fluid flow induced PGE2 release by bone cells is reduced by glycocalyx degradation whereas calcium signals are not, Biorheology 40, 591-603.
7. Yellowley, C.E., Li, Z., Zhou, Z., Jacobs, C.R., Donahue, H.J., 2000. Functional gap junctions between osteocytic and osteoblastic cells, J. Bone Miner. Res. 15, 209-217.

8. Knothe Tate, M.L., Adamson, J.R., Tami, A.E., Bauer, T.W., 2004. The osteocyte, Int. J. Biochem. Cell Biol. 36, 1-8.
9. Kufahl, R.H., Saha, S., 1990. A theoretical model for stress-generated fluid flow in the canaliculi-lacunae network in bone tissue, J. Biomech. 23, 171-180.
10. Moss, M.L., 1997. The functional matrix hypothesis revisited. 2. The role of an osseous connected cellular network, Am. J. Orthod. Dentofacial Orthop. 112, 221-226.
11. Reilly, G.C., Knapp, H.F., Stemmer, A., Niederer, P., Knothe Tate, M.L., 2001. Investigation of the morphology of the lacunocanalicular system of cortical bone using atomic force microscopy, Ann. Biomed. Eng. 29, 1074-1081.
12. Ponik, S.M., Pavalko, F.M., 2004. Formation of focal adhesions on fibronectin promotes fluid shear stress induction of COX-2 and PGE2 release in MC3T3-E1 osteoblasts, J. Appl. Physiol. 97, 135-142.
13. Brock, A., Chang, E., Ho, C.C., Leduc, P., Jiang, X., Whitesides, G.M., Ingber, D.E., 2003. Geometric determinants of directional cell motility revealed using microcontact printing, Langmuir 19, 1611-1617.
14. Chen, C.S., Alonso, J.L., Ostuni, E., Whitesides, G.M., Ingber, D.E., 2003. Cell shape provides global control of focal adhesion assembly, Biochem. Biophys. Res. Com. 307, 355-361.
15. Chen, C.S., Mrksich, M., Huang, S., Whitesides, G.M., Ingber, D.E., 1998. Micropatterned surfaces for control of cell shape, position, and function, Biotechnol. Prog. 14, 356-363.
16. Mrksich, M., Dike, L.E., Tien, J., Ingber, D.E., Whitesides, G.M., 1997. Using microcontact printing to pattern the attachment of mammalian cells to self-assembled monolayers of alkanethiolates on transparent films of gold and silver, Exp. Cell Res. 235, 305-313.
17. Singhvi, R., Kumar, A., Lopez, G.P., Stephanopoulos, G.N., Wang, D.I., Whitesides, G.M., Ingber, D.E., 1994. Engineering cell shape and function, Science 264, 696-698.
18. Healy, K.E., Thomas, C.H., Rezania, A., Kim, J.E., Mckeown, P.J., Lom, B., Hockberger, P.E., 1996. Kinetics of bone cell organization and mineralization on materials with patterned surface chemistry, Biomaterials 17, 195-208.
19. Sorribas, H., Padeste, C., Tiefenauer, L., 2002. Photolithographic generation of protein micropatterns for neuron culture applications, Biomaterials 23, 893-900.
20. Prime, K.L., Whitesides, G.M., 1991. Self-assembled organic monolayers: model systems for studying adsorption of proteins at surfaces, Science 252, 1164-1167.
21. Chen, C.S., Ostuni, E., Whitesides, G.M., Ingber, D.E., 2000. Using self-assembled monolayers to pattern ECM proteins and cells on substrates, Methods Molec. Bio. 139, 209-219.
22. Jiang, X., Ferrigno, R., Mrksich, M., Whitesides, G.M., 2003. Electrochemical desorption of self-assembled monolayers noninvasively releases patterned cells from geometrical confinements, J. Am. Chem. Soc. 125, 2366-2367.

23. Love, J.C., Wolfe, D.B., Chabinyc, M.L., Paul, K.E., Whitesides, G.M., 2002. Self-assembled monolayers of alkanethiolates on palladium are good etch resists, J. Am. Chem. Soc. 124, 1576-1577.
24. Hasenbein, M.E., Andersen, T.T., Bizios, R., 2002. Micropatterned surfaces modified with select peptides promote exclusive interactions with osteoblasts, Biomaterials 23, 3937-3942.
25. Feldkamp, L.A., Goldstein, S.A., Parfitt, A.M., Jesion, G., Kleerekoper, M., 1989. The direct examination of three-dimensional bone architecture in vitro by computed tomography, J. Bone Miner. Res. 4, 3-11.
26. Goldberg, G.S., Bechberger, J.F., Naus, C.C., 1995. A pre-loading method of evaluating gap junctional communication by fluorescent dye transfer, Biotechniques 18, 490-497.
27. Hantschel, T., Chow, E.M., Rudolph, D., Fork, D.K., 2002. Stressed metal probes for atomic force microscopy, Appl. Phys. Lett. 81, 3070-3072.
28. Hantschel, T., Chow, E.M., Rudolph, D., Fork, D.K., 2003. Stessed-metal NiZr probes for atomic force microscopy, Microelectron Eng. 67-68, 803-809.
29. Chow, E.M., Chua, C., Hantschel, T., Van Schuylenbergh, K., Fork, D.K., 2006. Pressure contact micro-springs in small pitch flip-chip packages, IEEE Trans. Comp. Packag. Tech. 29, in press (doi: 10.1109/TCAPT.2006.885959).
30. Donahue, S.W., Donahue, H.J., Jacobs, C.R., 2003. Osteoblastic cells have refractory periods for fluid-flow-induced intracellular calcium oscillations for short bouts of flow and display multiple low-magnitude oscillations during long-term flow, J. Biomech. 36, 35-43.
31. Bailey, C.H., Kandel, E.R., 1993. Structural changes accompanying memory storage, Annu. Rev. Physiol. 55, 397-426.
32. Turner, C.H., Robling, A.G., Duncan, R.L., Burr, D.B., 2002. Do bone cells behave like a neuronal network? Calcif. Tissue Int. 70, 435-442.
33. Bliss, T.V., Gardner-Medwin, A.R., 1973. Long-lasting potentiation of synaptic transmission in the dentate area of the unanaestetized rabbit following stimulation of the perforant path, J. Physiol. 232, 357-374.
34. Hebb, D.O., 1949. The organization of behavior; a neuropsychological theory, New York: Wiley.
35. Xia, S.L., Ferrier, J., 1992. Propagation of a calcium pulse between osteoblastic cells, Biochem. Biophys. Res. Commun. 186, 1212-1219.
36. Jorgensen, N.R., Geist, S.T., Civitelli, R., Steinberg, T.H., 1997. ATP- and gap junction-dependent intercellular calcium signaling in osteoblastic cells, J. Cell Biol. 139, 497-506.
37. Romanello, M., D'andrea, P., 2001. Dual mechanism of intercellular communication in HOBIT osteoblastic cells: a role for gap-junctional hemichannels, J. Bone Miner. Res. 16, 1465-1476.
38. Hung, C.T., Pollack, S.R., Reilly, T.M., Brighton, C.T., 1995. Real-time calcium response of cultured bone cells to fluid flow, Clin. Orthop. Relat. Res. 313, 256-269.

39. Tomkinson, A., Reeve, J., Shaw, R.W., Noble, B.S., 1997. The death of osteocytes via apoptosis accompanies estrogen withdrawal in human bone, J. Clin. Endocrinol. Metab. 82, 3128-3135.
40. Verborgt, O., Gibson, G.J., Schaffler, M.B., 2000. Loss of osteocyte integrity in association with microdamage and bone remodeling after fatigue in vivo, J. Bone Miner. Res. 15, 60-67.
41. Rubin, C.T., Bain, S.D., Mcleod, K.J., 1992. Suppression of the osteogenic response in the aging skeleton, Calcif. Tissue Int. 50, 306-313.
42. Turner, C.H., Takano, Y., Owan, I., 1995. Aging changes mechanical loading thresholds for bone formation in rats, J. Bone Miner. Res. 10, 1544-1549.
43. Takai, E., Mauck, R.L., Hung, C.T., Guo, X.E., 2004. Osteocyte viability and regulation of osteoblast function in a 3D trabecular bone explant under dynamic hydrostatic pressure, J. Bone Miner. Res. 19, 1403-1410.
44. Doty, S.B., 1981. Morphological evidence of gap junctions between bone cells, Calcif. Tissue Int. 33, 509-512.
45. Takayama, S., Mcdonald, J.C., Ostuni, E., Liang, M.N., Kenis, P.J., Ismagilov, R.F., Whitesides, G.M., 1999. Patterning cells and their environments using multiple laminar fluid flows in capillary networks, Proc. Natl. Acad. Sci. USA 96, 5545-5548.

INTRACELLULAR MEASUREMENTS OF STRAIN TRANSFER WITH TEXTURE CORRELATION

C. L. GILCHRIST, F. GUILAK AND L. A. SETTON

Departments of Biomedical Engineering and Orthopaedic Surgery
Duke University, Durham, NC, USA
E-mail: setton@duke.edu

The mechanisms by which cells sense and interpret mechanical signals may depend on how extracellular stimuli are transferred to and distributed within the cell. The magnitude or subcellular localization of a signal, such as mechanical deformation, may yield a specific cellular response. Examining how external stimuli are transferred to cells and the factors influencing this transfer may yield insight into the mechanisms underlying cellular mechanotransduction. In this study, we utilize fluorescent intracellular labels and an image analysis method, texture correlation, to study the transfer of extracellular mechanical deformations from an underlying substrate to specific regions and organelles within attached cells. Cells isolated from intervertebral disc tissues were cultured on deformable, elastic substrates and subjected to varying levels of tensile stretch, measuring the amount of strain transferred from substrate to cell cytoplasm and nucleus. Results indicate that a significant portion of strain was transferred from the substrate to cell cytoplasm, with an average strain transfer ratio (STR) of 0.79. In contrast, strain transferred to the nucleus was found to be much lower (STR = 0.17). Additionally, levels of strain transfer were found to depend upon a cell's alignment with the direction of applied stretch. This study presents a novel method for investigations of cellular mechanics, including measurements of intranuclear strains, with findings of differential magnitudes and patterns of strain transferred from the substrate to cell cytoplasm and nucleus.

1 Introduction

Cells in the body are subjected to a variety of mechanical stimuli including fluid flows, hydrostatic pressures, and mechanical deformations. These mechanical stimuli have been shown to play roles in regulating diverse cellular processes that include cell survival, differentiation, proliferation, and motility, for a variety of cell types [1-4]. The means by which cells sense these extracellular stimuli and convert them into intracellular biochemical responses, known as "mechanotransduction," may involve cellular machinery located in particular spatial regions within the cell [5]. At the cell surface, receptors such as integrins [6] and mechanosensitive ion channels [7] are known to play roles in the sensing of mechanical stimuli. Mechanical deformations or stresses may also be transmitted intracellularly to various adapter proteins (e.g. focal adhesions) [8,9], organelles, and the cytoskeleton [10]. Additionally, external stimuli may be transferred across the cytoplasm to the cell nucleus, where mechanical deformations or stresses have been hypothesized to alter gene expression [11,12]. Thus, the spatial localization (and magnitude) of mechanical stimuli transferred to regions within the cell may result in a specific cellular response.

Several approaches have been utilized by investigators to study how cells transfer and distribute externally applied mechanical stimuli to regions within the cell. Ligand-coated magnetic beads have been used to apply point-loads or moments to the apical surface of adherent cells, with measurements of cytoplasmic and nuclear displacements that suggest cells may transfer these stimuli significant distances away from the point of application [11,13]. Cells transfected with GFP-labeled intermediate filaments have been utilized to map cytoskeletal deformations in response to fluid shear [14]. Another approach used to study transfer of mechanical stimuli is to examine intracellular deformations of adherent cells in response to deformation of an underlying substrate. Several studies have utilized fluorescently-labeled beads (injected intracellularly) [15] or dyes [16] to track cytoplasmic deformations of cells attached to elastic deformable substrates. These methods allowed estimation of the stretch transferred from the substrate to the cell; however, strain estimates were only possible over large areas of the cytoplasm (~25-100 μm^2).

In a recent study [17], we have applied confocal microscopy and an image analysis method, texture correlation, to measure the transfer of strain from an underlying deformable substrate to regions within the cytoplasm and nuclei of living cells. Fibroblast-like cells isolated from the anulus region of intervertebral discs were cultured on deformable substrates and labeled with fluorescent dyes specific for cell mitochondria and nuclei for visualization with confocal microscopy. The substrate was subjected to varying levels of uniaxial tensile stretch, and high-resolution strains (2 μm resolution) were measured within the cell cytoplasm and nucleus and compared to those of the underlying substrate. We tested the hypothesis that strains would be transferred fully from underlying substrate to cell cytoplasm and nuclei, and that the amount of strain transferred to the cell would depend upon the cell's alignment with the direction of applied stretch. The method presented here may be applied to examine the roles of specific extracellular factors on the levels and spatial localization of transfer of mechanical stimuli.

2 Materials and Methods

2.1 Primary cell isolation and culture

Cells from the anulus fibrosus region of freshly harvested porcine intervertebral discs were isolated via enzymatic digestion [18] and cultured (37°C, 5% CO_2) in sub-confluent monolayers for 4-10 days (F-12 medium with 10% FBS, 10 mM HEPES, antibiotics). These cells exhibit a fibroblast-like phenotype, as indicated by their elongated cell morphology *in vitro* and *in vivo*, and synthesis of type I collagen [19,20].

Deformable elastic substrates consisted of silicone membranes (SILTEC gloss membrane, 35mm x 10mm x 0.25mm thickness, Technical Products) coated on one side with a thin layer of fluorescent microspheres (Fluospheres, Ex/Em: 535/575 nm, 2 μm diameter, Molecular Probes) suspended in silicone adhesive (GE RTV 108), providing markers for tracking substrate deformation, as shown in Figure 1. Membranes were sterilized under UV light (30 min.), and the opposite side was coated with type I bovine collagen (Sigma, 40 μg/mL in PBS, overnight at 4°C) to promote cell attachment. Membranes were then seeded with 10,000 cells/cm^2 and cultured for 48h to allow for attachment. Cell nuclei and mitochondria were fluorescently labeled using SYTO 82 (10 μM, Ex/Em: 541/560 nm, Molecular Probes) and MitoTracker Deep Red 633 (1.5 μM, Ex/Em: 644/665 nm, Molecular Probes), respectively, for 40 minutes prior to experiments.

2.2 *Stretch experiments*

Uniaxial stretch was applied to the membranes using a micrometer-controlled displacement device (modified from [21]) mounted on the stage of an inverted confocal laser scanning microscope (Zeiss LSM 510, Carl Zeiss), as shown in Figure 1. A coverslip window in the bottom of the chamber allowed for visualization of attached cells (inferior surface) and membrane markers (superior surface). Undeformed (reference) images were first acquired of cell mitochondria and nuclei (63X water immersion objective, NA 1.2, 0.7X zoom, 206 μm x 206 μm image field, 0.2 μm/pixel resolution, 2.0 μm slice thickness for mitochondria, 4.0 μm thickness for nuclei), followed by adjusting the focus to image microsphere markers on the superior surface of the membrane (20X objective, NA 0.5, 460 μm x 460 μm image field, 0.45 μm/pixel). A uniaxial (x-direction) stretch was then applied to the membrane to achieve substrate stretch values between 1.05-1.15. Images of the deformed cells and substrate were obtained immediately following stretch.

2.3 *Displacement measurement and strain calculations*

Displacements of cells and substrate were measured using a two-dimensional texture correlation algorithm [22]. With this method, pixel displacements are determined by comparing intensity patterns between “reference” and “deformed” images [23-25]. A given pixel is identified by a square subset of surrounding pixels (subset mask, m x m pixels), and the pixel’s displaced position in the deformed image is determined using a correlation algorithm. The algorithm has been shown to provide displacement measurements with sub-pixel resolution. Displacement fields are then calculated using a bicubic smoothing spline (MATLAB function csaps.m, The Mathworks, Inc.) and differentiated to determine Lagrangian finite strains (E_{xx}, E_{yy}, E_{xy}).

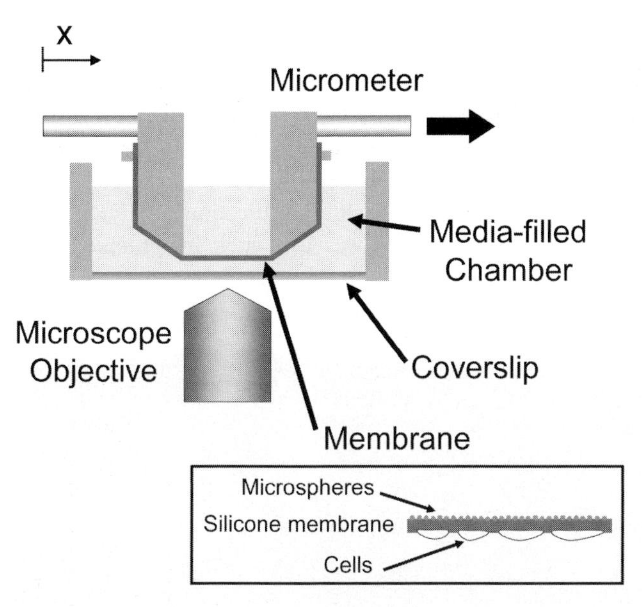

Figure 1. Micrometer device used to apply uniaxial stretch to cell-seeded silicone membranes. Membranes were inverted for viewing cells on their substrates through a coverslip.

2.4 Intracellular strain calculations following stretch

A series of stretch experiments was performed (56 cells analyzed in 10 independent experiments) with substrates subjected to stretches in the x-direction ranging from approximately 1.05 to 1.15. Cytoplasmic, nuclear, and substrate strain fields were determined using images of acquired before and immediately following substrate stretch. For cytoplasmic and nuclear strain calculations, a 4x4 grid of pixels (strain resolution of 2 μm) was selected for displacement tracking (16 measurement points, 10 pixel spacing between grid points chosen over a 36 μm^2 measurement area) contained entirely within the cell cytoplasm or nucleus, as shown in Figure 2. In calibration studies [17], strain error associated with this technique was shown to be less than 0.01 for both cytoplasmic and nuclear images. Substrate strain was calculated for each stretch experiment and averaged across an image field matched to the image field of the analyzed cells (270 μm x 270 μm area, 7x7 grid). For an individual cell, the strain transferred from the substrate to the cell was defined as the strain transfer ratio (STR), calculated as the ratio of the mean cell strain (cytoplasmic or nuclear) to the mean underlying substrate strain. STR values were compared to 1 (t-test, $p<0.05$ significant) to test the hypothesis that substrate strain was fully transferred to cell cytoplasm and nucleus. Pairwise comparisons of STR values were also made between cell region (cytoplasm versus nucleus) and direction (parallel and transverse to direction of applied stretch) for each cell (paired t-test,

Bonferroni correction). The dependence of strain transfer on level of strain magnitude was analyzed via linear regression.

Additionally, we tested for correlations of cell alignment with STR for cytoplasm and nuclei. The aspect ratio (cell length/width) of each cell was measured using image analysis software (Zeiss AIM), and the alignment angle θ for a cell was calculated as the angle between the x-direction and the direction of the cell's long axis, with an angle of 0° representing alignment with the direction of stretch. The dependence of STR on cell alignment (cos(θ)) was evaluated by grouping cells based on alignment and comparing via ANOVA.

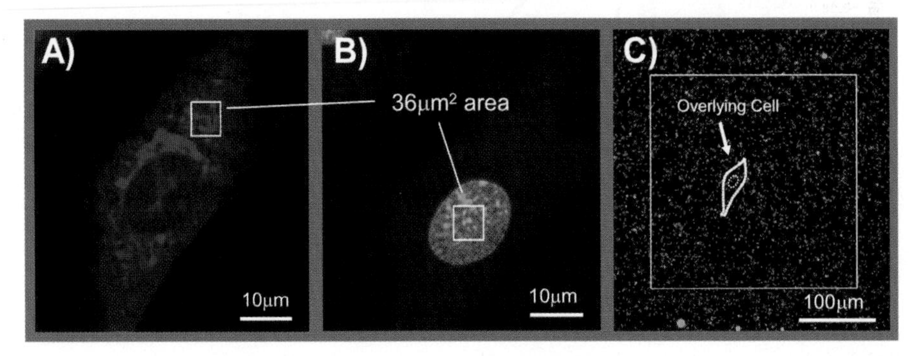

Figure 2. (A) Representative images of fluorescently labeled cell mitochondria showing square region within which strain was measured (white box, 36 μm^2). (B) Image of labeled cell nucleus from same cell. (C) Image of underlying substrate showing fluorescent microsphere markers and region of strain measurement (white box, 72,900 μm^2). An overlay of the cell corresponding to (A) and (B) is shown.

3 Results

3.1 Cell stretching experiments

A total of ten uniaxial stretch experiments were performed, with associated imposed substrate strains ranging from 0.045 to 0.14. Measured substrate strains for each experiment were found to be very uniform (e.g., Fig.3), with standard deviations in E_{xx} for the substrate typically less than 5% of the mean strain values (for example, $E_{xx} = 0.089 \pm 0.002$, Fig. 3A).

Mean strains within the cytoplasmic region of a cell were found to be typically similar to those of the underlying substrate, although significant heterogeneity was found within the cytoplasm of a given cell (standard deviations of 0.024 for E_{xx}, 0.014 for E_{yy}, and 0.012 for E_{xy} for cytoplasmic strains, for all cells). Significant variability was also noted between cells on the same substrate; as an example, for the experiment shown in Figure 3 mean E_{xx} cytoplasmic strains ranged from 57% to 112% of the mean substrate strain. For cytoplasmic strains in the direction of applied stretch (x-direction), the strain transfer ratio was found to be STR_x = 0.79±0.34, which was significantly less than 1 ($p<0.0001$), indicating cytoplasmic strains in the direction of applied stretch were not fully transferred. In the transverse direction (y-direction), STR_y= 0.99±0.68 was not found to be different from 1 ($p=0.88$).

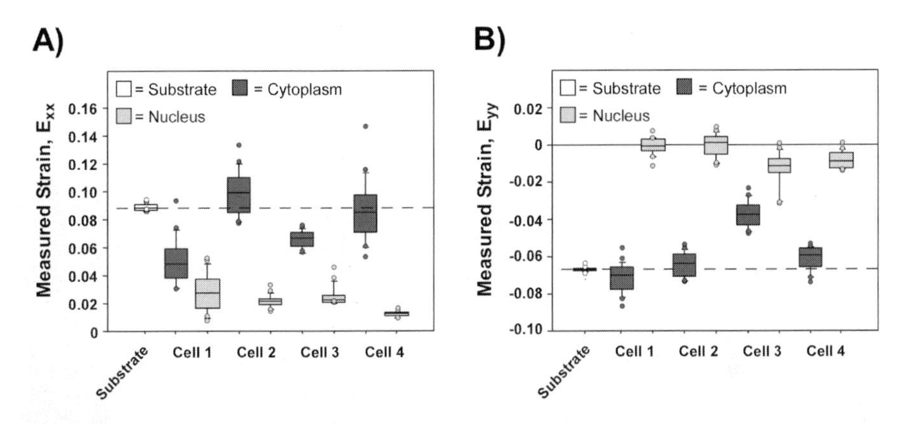

Figure 3. Measured (A) longitudinal (E_{xx}) and (B) transverse (E_{yy}) substrate (n=49 strain measurements per substrate) and cell (n=16 strain measurements per cell for cytoplasm and nucleus) strains for a representative uniaxial stretch experiment. For each cell, measured cytoplasmic (dark) and nuclear (light) strain measurements are shown. Box plot shows inter-quartile range (box) and median (line). Whiskers identify 90th percentile range, with data points falling outside of this range shown (o). Figure reprinted with permission from [17].

Strains within the cell nucleus were much lower than corresponding substrate and cytoplasmic strains, with STR_x = 0.17±0.28 and STR_y = 0.38±0.89. These values were significantly different from 0 and 1 ($p<0.0001$), and lower than those of the cell cytoplasm ($p<0.0001$). In contrast to cytoplasmic strains, nuclear strains were also found to be more homogeneous, with standard deviations less than 0.01 for all strain components.

Cytoplasmic strains were found to correlate with substrate strains along the direction of applied stretch, as shown in Figure 4 (slope = 1.02, $p<0.0001$, R^2 = 0.36). In contrast, nuclear strains were found to be only weakly correlated with those of the underlying substrate, (slope = 0.25, $p<0.0001$, R^2 = 0.08). For both cytoplasmic and nuclear correlations, linear-fits were found to have positive x-intercepts (as indicated by arrows in Fig.4; x=0.021 for cytoplasmic, x=0.032 for nuclear) that were significantly different from zero ($p<0.01$).

An analysis of variance detected differences in cytoplasmic STR_x between a group of cells "aligned" ($-45<\theta<+45$, n=24) with the direction of applied stretch and those that were "unaligned" ($\theta<-45$, $\theta>+45$, n=31), as shown in Figure 5 ($p=0.015$). Thus, cell alignment with the applied stretch direction resulted in higher strain transfers to the cell cytoplasm. In the direction transverse to applied stretch (STR_y), "unaligned" cells showed a trend towards higher STR_y as compared to "aligned cells," although these differences were not detected as significant (Fig. 5).

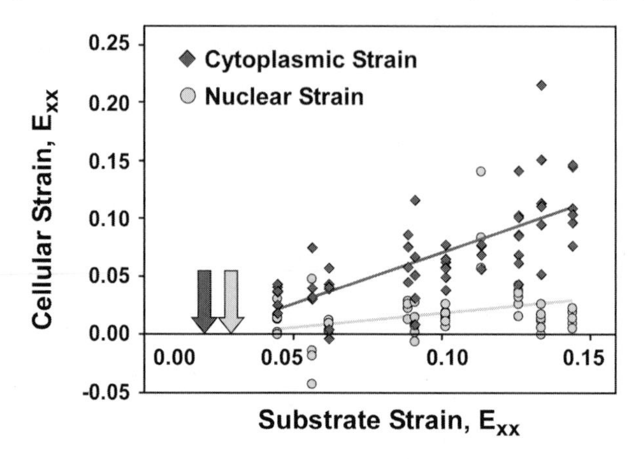

Figure 4. Measured mean substrate strain versus measured mean cytoplasmic and nuclear strains for cells subjected to uniaxial stretch (in x-direction). Data shown represent 10 separate stretch experiments and a total of 56 cells. Measured cytoplasmic strains (♦ represents mean within a single cell) were found to correspond with substrate strain (dark line, slope = 1.02, $p<0.0001$, $R^2 = 0.36$). Measured nuclear strains (○) for same cells were much lower but did correspond with substrate strain (light line, slope = 0.25, $p<0.0001$, $R^2 = 0.08$). Dark and light arrows indicate non-zero intercepts for linear fits of cytoplasm and nucleus, respectively. Figure reprinted with permission from [17].

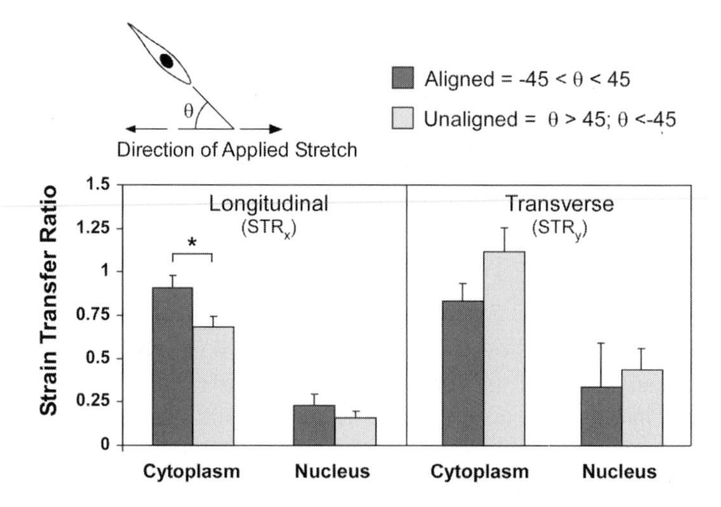

Figure 5. Effects of cell alignment on strain transfer for cells subjected to uniaxial stretch. Mean cytoplasmic and nuclear strain transfer ratios longitudinal (x-direction) and transverse (y-direction) to direction of applied substrate stretch for "aligned" (n=24) and "unaligned" groups, (n=31). * p=0.015, ANOVA. Figure reprinted with permission from [17].

4 Discussion

In this study, we present a novel method to quantify strain fields in cells at the subcellular level using fluorescent cell labeling and texture correlation image analysis. Strain transfer from underlying substrate to the cytoplasm and nucleus of attached cells was evaluated. The strain transfer ratio from substrate to cytoplasm was found to be significantly less than 1 (STR_x = 0.79 for all experiments), indicating that strains were not fully transferred to the cell. However, strain was transferred from substrate to cytoplasm with increasing stretch magnitude at an approximately 1:1 ratio (slope = 1.02), after an initial offset or non-linear region at low strain levels (Fig.4). The observed strain offset may reflect a passive, non-linear mechanical behavior or an active cell response following imposition of substrate stretch. In stark contrast to cytoplasmic strains, little strain was transferred from substrate to nucleus (STR_x =0.17) and was only slightly correlated with substrate strains (slope = 0.25). Similar to the observations for cytoplasmic strain, there was evidence of nuclear strain offset at 0.0 substrate strain.

The findings of this study showed that the magnitudes of nuclear strains were significantly smaller and relatively uniform as compared to the surrounding cytoplasm, for a given cell. These findings are consistent with previous reports suggesting the presence of a physical linkage that can transmit mechanical deformation from the substrate, across the cell membrane, to the cell nuclei (e.g. [11,15,26]). Importantly, nuclei experienced a relatively small fraction (<25%) of the cytoplasmic strain, consistent with a previous study showing that endothelial cell nuclei deform significantly less (50-80% less) than the underlying substrate [15]. Similarly, compression of chondrocytes within the native cartilage extracellular matrix showed that nuclear deformation was significantly less than that of the whole chondrocyte [27]. The difference between the magnitudes of nuclear and cytoplasmic strains may be due to the significantly higher mechanical stiffness of the nucleus [28-30]. Alternatively, the smaller nuclear strains may be due limited or indirect cytoskeletal connection between the extracellular substrate and nucleus. Of interest was the finding that the distribution of nuclear strain was significantly more uniform than the strains in the surrounding cytoplasm, with standard deviations of nuclear strain that were less than 50% of those in the cytoplasm. This finding may also reflect more homogeneous mechanical properties of the nucleus as compared to the cytoplasm, at the current scale of the strain measurements.

Cells that were aligned more highly with the direction of applied stretch were found to exhibit higher cytoplasmic strain transfer than those that were not aligned, suggesting that the coupling between a cell's cytoplasm and the underlying substrate is higher for these aligned cells. This finding could be a result of mechanical anisotropy of the cell's cytoplasm, which may correspond with the cell's long axis [31]. Measurement of cell features that may contribute to this anisotropy, including cytoskeletal architecture (e.g. actin stress fiber alignment, number) and focal adhesion densities or area, may provide insight into the mechanisms of anisotropic strain transfer. Utilization of cells expressing fluorescently-labeled cytoskeletal [14] or adhesion [32] proteins could facilitate direct examination of these factors on strain transfer behavior in individual cells.

Cytoplasmic strains measured in this study were found to be highly heterogeneous, both within individual cells and between cells on a given substrate. The variability in strains observed was much higher than that estimated in our error analyses, and thus likely represents actual heterogeneity within the cell's cytoplasm and differences in coupling between individual cells and their underlying substrate. Indeed, the resolution of the technique presented here may be high enough to detect localized differences in cytoskeletal architecture or organelle positioning. The observed intracellular and cell-substrate variabilities in strains may also be a result of active reorganization by the cell in response to substrate stretch, particularly for cells exhibited very high or low strain transfer behaviors.

The methods presented in this study have recently been extended to capture the response of single cells across varying levels of applied substrate strain, as shown for a single cell in Figure 6. In the experiment shown, a single cell was tracked at five levels of increasing substrate stretch, with strain transfer from substrate to cytoplasm linear across a range of substrate strains in the direction of applied stretch. In contrast, transfer to the nucleus initially remained close to zero but increased distinctly above a threshold substrate strain. Thus, examining single cell responses may provide more detailed understanding of strain transfer behaviors and reduce confounding effects of biological intercellular variability.

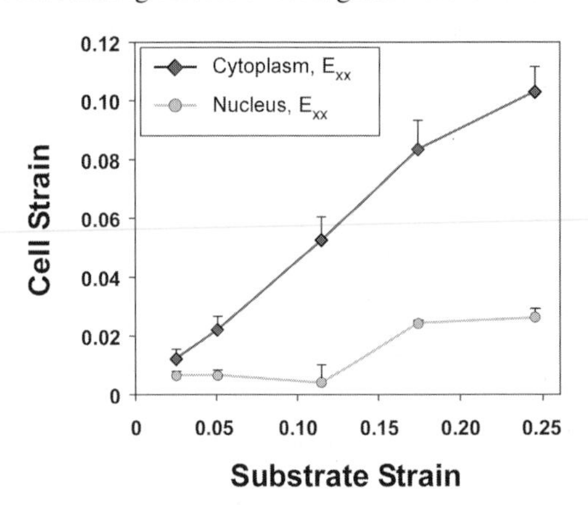

Figure 6. Measured cytoplasmic (dark) and nuclear (light) strains for a single cell subjected to increasing levels of uniaxial substrate stretch.

The technique presented here may be utilized to study mechanisms of strain transfer, including examining the roles of complex extracellular inputs on cellular deformation responses. Isolating the effects of individual extracellular inputs (i.e. "uncoupling" inputs) such as extracellular ligand, ligand density, receptor, and substrate compliance may provide insight into cell behavior in more complex *in situ* environments, and yield information for modeling of cellular mechanics within intact tissues.

Acknowledgments

The authors acknowledge the contributions of S. Witvoet-Braam to this work. The work was supported by NIH grants AR47442, AG15768, AR48182, AR50245 and T32-GM08555.

References

1. Estes, B.T., Gimble, J.M., Guilak, F., 2004. Mechanical signals as regulators of stem cell fate, Curr. Top. Dev. Biol. 60, 91-126.
2. Guilak, F., Sah, R.L., Setton, L.A., 1997. Physical regulation of cartilage metabolism, In: Mow, V.C., Hayes, W.C. (Eds.), Basic Orthopaedic Biomechanics, Lippincott-Raven, Philadelphia, pp. 179-207.
3. Hsieh, M.H., Nguyen, H.T., 2005. Molecular mechanism of apoptosis induced by mechanical forces, Int. Rev. Cytol. 245, 45-90.
4. Li, Y.S., Haga, J.H., Chien, S., 2005. Molecular basis of the effects of shear stress on vascular endothelial cells, J. Biomech. 38, 1949-1971.
5. Iqbal, J., Zaidi, M., 2005. Molecular regulation of mechanotransduction, Biochem. Biophys. Res. Commun. 328, 751-755.
6. Katsumi, A., Orr, A.W., Tzima, E., Schwartz, M.A., 2004. Integrins in mechanotransduction, J. Biol. Chem. 279, 12001-12004.
7. Sachs, F., Morris, C.E., 1998. Mechanosensitive ion channels in nonspecialized cells, Rev. Physiol. Biochem. Pharmacol. 132, 1-77.
8. Shemesh, T., Geiger, B., Bershadsky, A.D., Kozlov, M.M., 2005. Focal adhesions as mechanosensors: a physical mechanism, Proc. Natl. Acad. Sci. USA 102, 12383-12388.
9. Bershadsky, A.D., Balaban, N.Q., Geiger, B., 2003. Adhesion-dependent cell mechanosensitivity, Annu. Rev. Cell Dev. Biol. 19, 677-695.
10. Wang, N., Butler, J.P., Ingber, D.E., 1993. Mechanotransduction across the cell surface and through the cytoskeleton, Science 260, 1124-1127.
11. Maniotis, A.J., Chen, C.S., Ingber, D.E., 1997. Demonstration of mechanical connections between integrins cytoskeletal filaments, and nucleoplasm that stabilize nuclear structure, Proc. Natl. Acad. Sci. USA 94, 849-854.
12. Ingber, D.E., Prusty, D., Sun, Z., Betensky, H., Wang, N., 1995. Cell shape, cytoskeletal mechanics, and cell cycle control in angiogenesis, J. Biomech. 28, 1471-1484.
13. Hu, S.H., Chen, J.X., Fabry, B., Numaguchi, Y., Gouldstone, A., Ingber, D.E., Fredberg, J.J., Butler, J.P., Wang, N., 2003. Intracellular stress tomography

reveals stress focusing and structural anisotropy in cytoskeleton of living cells, Am. J. Physiol. Cell Physiol. 285, C1082-C1090.
14. Helmke, B.P., Rosen, A.B., Davies, P.F., 2003. Mapping mechanical strain of an endogenous cytoskeletal network in living endothelial cells, Biophys. J. 84, 2691-2699.
15. Caille, N., Tardy, Y., Meister, J.J., 1998. Assessment of strain field in endothelial cells subjected to uniaxial deformation of their substrate, Ann. Biomed. Eng. 26, 409-416.
16. Wall, M.E., Weinhold, P.S., Siu, T., Brown, T.D., Banes, A.J., 2007. Comparison of cellular strain with applied substrate strain in vitro, J. Biomech. 40, 173-181.
17. Gilchrist, C.L., Witvoet-Braam, S.W., Guilak, F., Setton, L.A., 2007. Measurement of intracellular strain on deformable substrates with texture correlation, J. Biomech. in press (doi: 10.1016/j.jbiomech.2006.1003.1013).
18. Baer, A.E., Wang, J.Y., Kraus, V.B., Setton, L.A., 2001. Collagen gene expression and mechanical properties of intervertebral disc cell-alginate cultures, J. Orthop. Res. 19, 2-10.
19. Horner, H.A., Roberts, S., Bielby, R.C., Menage, J., Evans, H., Urban, J.P., 2002. Cells from different regions of the intervertebral disc: effect of culture system on matrix expression and cell phenotype, Spine 27, 1018-1028.
20. Wang, J.Y., Baer, A.E., Kraus, V.B., Setton, L.A., 2001. Intervertebral disc cells exhibit differences in gene expression in alginate and monolayer culture, Spine 26, 1747-1751.
21. Guilak, F., Ratcliffe, A., Mow, V.C., 1995. Chondrocyte deformation and local tissue strain in articular cartilage: a confocal microscopy study, J. Orthop. Res. 13, 410-421.
22. Gilchrist, C.L., Xia, J.Q., Setton, L.A., Hsu, E.W., 2004. High-resolution determination of soft tissue deformations using MRI and first-order texture correlation, IEEE Trans. Med. Imaging 23, 546-553.
23. Bay, B.K., 1995. Texture correlation: a method for the measurement of detailed strain distributions within trabecular bone, J. Orthop. Res. 13, 258-267.
24. Bey, M.J., Song, H.K., Wehrli, F.W., Soslowsky, L.J., 2002. A noncontact, nondestructive method for quantifying intratissue deformations and strains, J. Biomech. Eng. 124, 253-258.
25. Wang, C.C., Chahine, N.O., Hung, C.T., Ateshian, G.A., 2003. Optical determination of anisotropic material properties of bovine articular cartilage in compression, J. Biomech. 36, 339-353.
26. Hu, S., Chen, J., Butler, J.P., Wang, N., 2005. Prestress mediates force propagation into the nucleus, Biochem. Biophys. Res. Commun. 329, 423-428.
27. Guilak, F., 1995. Compression-induced changes in the shape and volume of the chondrocyte nucleus, J. Biomech. 28, 1529-1541.

28. Guilak, F., Tedrow, J.R., Burgkart, R., 2000. Viscoelastic properties of the cell nucleus, Biochem. Biophys. Res. Commun. 269, 781-786.
29. Caille, N., Thoumine, O., Tardy, Y., Meister, J.J., 2002. Contribution of the nucleus to the mechanical properties of endothelial cells, J. Biomech. 35, 177-187.
30. Dong, C., Skalak, R., Sung, K.L., 1991. Cytoplasmic rheology of passive neutrophils, Biorheology 28, 557-567.
31. Hu, S.H., Eberhard, L., Chen, J.X., Love, J.C., Butler, J.P., Fredberg, J.J., Whitesides, G.M., Wang, N., 2004. Mechanical anisotropy of adherent cells probed by a three-dimensional magnetic twisting device, Am. J. Physiol. Cell Physiol. 287, C1184-C1191.
32. Balaban, N.Q., Schwarz, U.S., Riveline, D., Goichberg, P., Tzur, G., Sabanay, I., Mahalu, D., Safran, S., Bershadsky, A., Addadi, L., Geiger, B., 2001. Force and focal adhesion assembly: a close relationship studied using elastic micropatterned substrates, Nat. Cell Biol. 3, 466-472.

II. CELL RESPONSE TO MECHANICAL STIMULATION

IDENTIFYING THE MECHANISMS OF FLOW-ENHANCED CELL ADHESION VIA DIMENSIONAL ANALYSIS

C. ZHU AND V. I. ZARNITSYNA

Coulter Department of Biomedical Engineering, and Woodruff School of Mechanical Engineering, Georgia Institute of Technology, Atlanta, GA 30332, USA
E-mail: cheng.zhu@bme.gatech.edu

T. YAGO AND R. P. MCEVER

Cardiovascular Biology Research Program, Oklahoma Medical Research Foundation, and Department of Biochemistry and Molecular Biology, University of Oklahoma Health Sciences Center, Oklahoma City, OK 73104, USA

Cell adhesion is mediated by specific receptor-ligand bonds. In several biological systems, increasing flow has been observed to enhance cell adhesion despite the increasing dislodging fluid shear forces. Flow-enhanced cell adhesion includes several aspects: flow augments the initial tethering to stationary surface of flowing cells, slows the velocity and increases the regularity of rolling cells, and increases the number of rollingly adherent cells. Mechanisms for this intriguing phenomenon may include transport-dependent acceleration of bond formation and force-dependent deceleration of bond dissociation. The former includes three distinct transport modes: sliding of cell bottom on the surface, Brownian motion of the cell, and rotational diffusion of the interacting molecules. The latter involves a recently demonstrated counterintuitive behavior called catch bonds where force prolongs rather than shortens the lifetimes of receptor-ligand bonds. In this article, we summarize our recently published data that used dimensional analysis to elucidate the above mechanisms for flow-enhanced leukocyte adhesion mediated by L-selectin-ligand interactions.

1 Introduction

Adhesion of blood cells to vascular surfaces occurs in a varying hydrodynamic environment of the circulation. Paradoxically, flow enhances adhesion in some systems. For example, leukocytes require a threshold shear to tether to and roll on endothelial cells at sites of inflammation and injury [1, 2]. Several aspects of flow-enhanced leukocyte adhesion are illustrated in Fig. 1. As wall shear stress drops below the threshold, fewer cells tether (Fig. 1A) and roll (Fig. 1B), and those that do roll more rapidly (Fig. 1C) and less regularly and detach more easily, which also reduces the number of rollingly adherent cells (Fig. 1B). These are counterintuitive because the higher the wall shear stress, the larger the dislodging forces exerted on the cells by the fluid. After reaching an optimal level (~ 1 dyn/cm^2 in Fig. 1), further increase in wall shear stress results in a decrease in tether rate (Fig. 1A), a decrease in the number of rollingly adherent cells (Fig. 1B), and an increase in rolling velocity (Fig. 1C), which are intuitive. As another example, platelets require a minimum flow to tether to and roll on the extracellular matrix of the vessel wall exposed after rupture of the endothelium lining [3, 4]. In addition, some enteric bacteria require a minimum flow to adhere to intestinal epithelia [5].

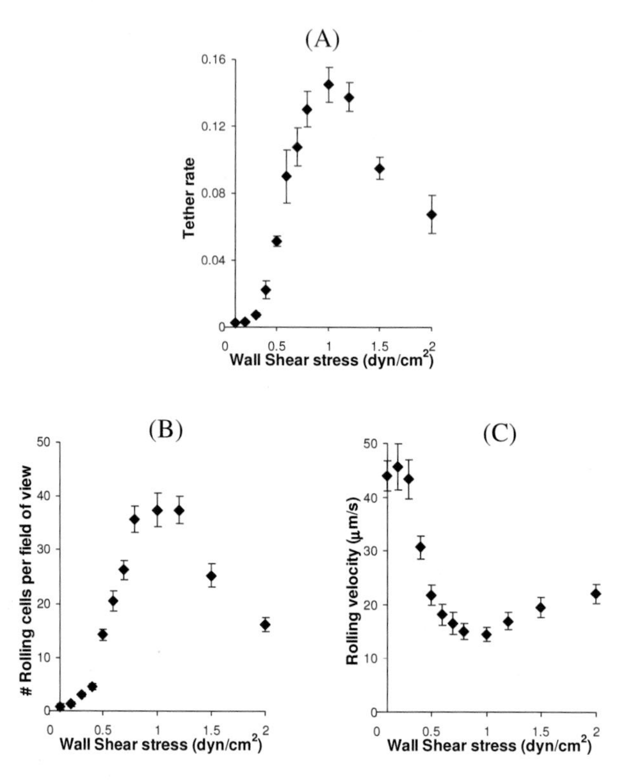

Figure 1. Increasing wall shear stress in a flow chamber initially enhances L-selectin-mediated neutrophil tethering (A) and rolling adhesion (B) to a surface coated with P-selectin glycoprotein ligand 1 (PSGL-1) and reduces the neutrophil rolling velocity on PSGL-1 (C). After reaching an optimal level (~ 1 dyn/cm^2), further increase in wall shear stress decreases neutrophil tethering (A) and rollingly adhesion (B) to PSGL-1 and increases the neutrophil rolling velocity on PSGL-1 (C). Data are presented as mean ± s.e.m.

Cell adhesion is mediated by specific interactions between adhesive receptors and ligands on the cell surface. In the above examples, leukocyte adhesion to vascular endothelium is mediated by interactions of selectins with glycoconjugates, e.g., P-selectin glycoprotein ligand 1 (PSGL-1). Platelet adhesion to damaged vessel wall is mediated by interactions of platelet glycoprotein Ib (GPIb) with von Willebrand factor (VWF). Bacterium adhesion to intestinal epithelium is mediated by interactions of FimH receptor with mannosylated glycoproteins. The interplay of these molecular interactions with flow is primarily responsible for the counterintuitive flow-enhanced cell adhesion, since flow-enhanced adhesion can be reconstituted in cell-free systems using purified adhesive receptors and ligands.

Receptor-ligand interactions are governed by the kinetics of rapid transition between bound and unbound states. Therefore, flow most likely exerts its effects through alteration of these kinetic rates. In the following sections, we will first show that flow-enhanced cell tethering (cf. Fig. 1A) is due to the interplay of transport and the kinetic on-rate (k_{on}) for the formation of receptor-ligand bonds. We will then show that flow-enhanced rolling (cf. Fig. 1C) is due to an unusual force regulation of the kinetic off-rate (k_{off}) for the dissociation of receptor-ligand bonds. Both of these mechanisms contribute to the flow-enhanced accumulation of rollingly adherent cells (cf. Fig. 1B).

2 Transport Governs Flow-enhanced Cell Tethering

2.1 Conceptual scheme of tethering process

For a cell (or microsphere) that moves with flowing blood to tether, its receptors must contact ligands on the vascular surface (or flow chamber floor in our experiments) for a sufficient duration. A contact requires a sufficiently small gap distance between the cell bottom and the surface for a receptor to reach a ligand. Tethering is a result of bond formation between the cell and the surface, which should be proportional to the densities of receptors (m_r) and ligands (m_l) as well as the contact area according to the law of mass action. The contact area A_i is proportional to the cell radius r and the combined length of the interacting molecules l_m in excess of the gap distance l_i (Fig. 2A). The cell is subjected to Brownian motion that randomly modulates the gap distance above and below the contact threshold $z = l_m$, breaking the observation time into alternating periods of brief contact (t_i) and noncontact (t_j) (Fig. 2B). The receptors and ligands are also subjected to rotational diffusion, which orients their binding sites for molecular docking (Fig. 2B). Hydrodynamic theory for motion of a sphere near a wall [6, 7] predicts that the angular velocity Ω of the sphere rotation cannot keep up with the translational velocity V of its center for pure rolling, such that the sphere bottom slides on the wall with a sliding velocity $V_s \equiv V - r\Omega$ proportional to $r\dot{\gamma}$ where $\dot{\gamma}$ is the wall shear rate (Fig. 2A). It has been shown that the probability P_a for a cell to tether before it flows a distance x over the surface is [8]:

$$P_a = 1 - \exp(-p_{ad}x) \tag{1}$$

where the probability p_{ad} for a cell to tether per unit distance is [9]:

$$p_{ad} = 2\pi m_r m_l r l_c t_c \phi k_{on}/V = -\ln(1 - \text{tether rate})/L \tag{2}$$

where ϕ is the collision frequency and $2\pi r l_c t_c = \langle A_i t_i \rangle$ is cross-correlation of contact area and contact time. Tether rate is defined as the ratio of the number of tethering events to the total number of cells flowing through the field of view of length L in a given period (e.g., one minute).

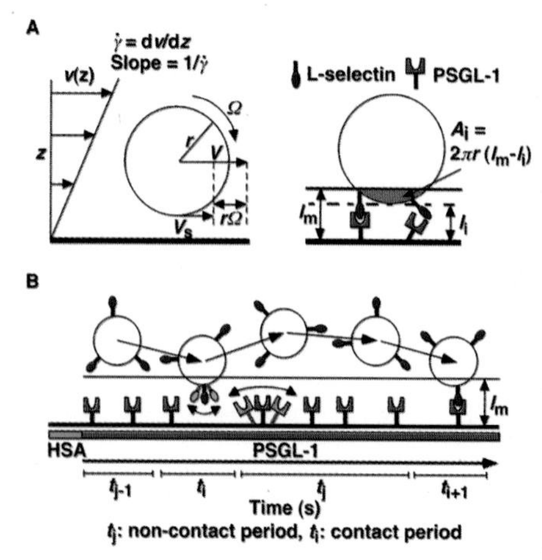

Figure 2. Parameters of cell tethering under flow. A, The fluid velocity v of a Couette flow field is parallel to the surface and increases linearly with the distance from the surface (z direction). The shear rate $\dot{\gamma} = dv/dz$. The sphere bottom has a positive velocity $V_s \equiv V - r\Omega$ where V is translational velocity of the sphere center, Ω is angular velocity of the sphere, and r is the sphere radius. The sphere and the surface are respectively coated with receptors and ligands whose combined length l_m sets a contact threshold. When the gap distance l_i between the sphere bottom and the surface is less than l_m, the two are in contact with an area A_i. B, The sphere is susceptible to thermal excitations that cause Brownian motion. This produces fluctuations in sphere z-position, which are depicted by the wavy trajectory of the sphere shown in five different times and positions. By randomly modulating the gap distance above and below the threshold for contact $Z = l_m$ (horizontal line), discontinuous contacts of different portions of the sphere with different portions of the surface are observed with alternating intervals of contact (t_i) and noncontact (t_j). A productive contact results in a tethering event, but many contacts are nonproductive. As schematically shown for one receptor and one ligand by the movements along the two-sided arrows (depicted by lighter colors), the binding sites of L-selectin and PSGL-1 can undergo rotational diffusion even though portions of the molecules are anchored to the respective sphere surface and chamber floor. To ensure that only first-time tethering events were observed, the chamber floor upstream to the microscope field of view was coated with HSA to allow measurement of the distance traveled by the sphere from the demarcation line to the location where tethering occurs. The cell, contact area and molecular sizes are not drawn to scale. Reproduced from Ref. [9].

Conceptually, bond formation can be divided into two steps: transport that brings two molecules in close proximity and reaction during which the interacting molecules dock. Depending on the relative time scales of the two steps, tethering can be transport-limited or reaction-limited. A faster transport produces more frequent collisions but also shortens the contact durations, which decreases cell tethering in the reaction-limited regime. Eq. 2 will serve as a guide to experiments that examine how the adhesion probability per distance p_{ad} depends on various transport mechanisms.

2.2 *Enhancing tethering by mean sliding velocity*

To apply dimensional analysis to identify transport mechanisms of cell tethering, we independently varied parameters that affect these mechanisms. We perfused cells or microspheres of different radii in media of different viscosities through a flow chamber at different wall shear rates. According to Eq. 2, the probability adhesion per distance p_{ad} could be normalized by dividing by $m_r m_l r$ to remove the mass action effect. $p_{ad}/(m_r m_l r)$ was then plotted against different parameters to dissect which mechanism dominated which regime of the curve (Fig. 3). Alignment of the entire curves or portions thereof when plotted against a parameter suggests the dominance of the mechanism controlled by this parameter in the regime where curves align.

Lack of alignment suggests the presence of competing mechanisms. Lack of sensitivity to the change of a parameter indicates its irrelevance.

As flow increased, the tether rates, measured by the normalized probability of adhesion per distance, $p_{ad}/(m_r m_l r)$, of L-selectin-bearing microspheres (Fig. 3, A-D) or neutrophils (Fig. 3, E and F) to PSGL-1 increased initially, reached a maximum, and then decreased. When plotted against the wall shear rate $\dot{\gamma}$, initial portions of the ascending phase of curves for different medium viscosities μ aligned for microspheres of the same radius r (but shifted with r) (Fig. 3A) and for neutrophils (Fig. 3E), suggesting two competing mechanisms respectively governed by $\dot{\gamma}$ and r. When the abscissa was rescaled by multiplying $\dot{\gamma}$ by r, the initial portions of the ascending phase of all microsphere curves collapsed (Fig. 3C), indicating that the product $r\dot{\gamma}$ is the governing parameter in this regime. The sliding velocity of the sphere bottom relative to the chamber floor is proportional to $r\dot{\gamma}$. Therefore, these results demonstrate that convective transport of receptors to ligands by relative sliding between the sphere bottom (where the receptors reside) and the surface (where the ligands reside) provides a dominant transport mechanism to enhance tethering when $r\dot{\gamma}$ is small.

2.3 Enhancing tethering by Brownian motion

Sliding could enhance tethering because bond formation was transport-limited. The alignment of the $p_{ad}/(m_r m_l r)$ vs. $r\dot{\gamma}$ curves broke down when $p_{ad}/(m_r m_l r)$ reached a maximum, suggesting that another transport mechanism limited tethering here. The $r\dot{\gamma}$ value where $p_{ad}/(m_r m_l r)$ achieved maximum (referred to as optimal $r\dot{\gamma}$) decreased with increasing sphere radius r and/or medium viscosity μ for both microspheres (Fig. 3C) and neutrophils (Fig. 3E). When plotted against D_s, the optimal $r\dot{\gamma}$ aligned into two nearly straight lines, one for microspheres and the other for neutrophils, regardless of the r and/or μ values (Fig. 4A). $D_s \equiv k_B T/(6\pi\mu r)$ (where k_B is the Boltzmann constant and T is absolute temperature) is the sphere diffusivity according to the Stokes-Einstein relationship, which can be used as a metric for the sphere Brownian motion. These results identify sphere Brownian motion as another transport mechanism that enhances tethering. This is intuitive as more vibrant Brownian motion should produce more frequent collisions with larger contact areas, expanding the capacity for sliding to further enhance tethering.

When $p_{ad}/(m_r m_l r)$ was plotted against the wall shear stress σ, the final portions of the descending phase of different curves aligned for microspheres of the same radius r (but shifted with r) (Fig. 3B) and for neutrophils (Fig. 3F). When the abscissa was rescaled by multiplying σ by $13.2r^2$, the final portions of the descending phase of all microsphere curves aligned even for those of different radii (Fig. 3D). $[F_t]_{max} \equiv 13.2r^2\sigma$ equals the force that tethers the sphere [6]. It also equals $13.2(k_B T/6\pi)(r\dot{\gamma}/D_s)$. Since $13.2(k_B T/6\pi)$ is a constant, this suggests that $r\dot{\gamma}/D_s$ is the governing parameter in this regime. This is intuitive because the larger the D_s, the further can $r\dot{\gamma}$ enhance tethering. Scaling $r\dot{\gamma}$ by D_s normalizes the capacity for sliding to enhance tethering, thereby aligning the curves.

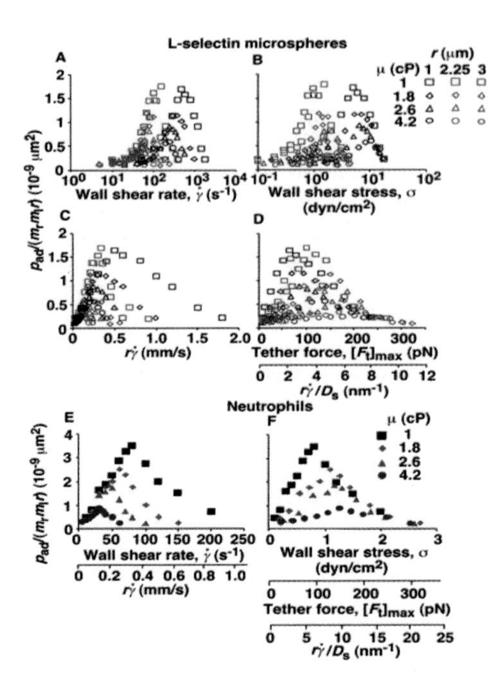

Figure 3. Adhesion probabilities per distance, p_{ad}, of L-selectin bearing microspheres (A-D) and neutrophils (E and F) were calculated from the tether rate data from Eq. 2, normalized by dividing by m_rm_lr, and plotted vs. wall shear rate $\dot{\gamma}$ (A and E), wall shear stress $\sigma = \mu\dot{\gamma}$ (B and F), product $r\dot{\gamma}$ (C and E), and maximum tether force $[F_t]_{max}$ = 13.2 $r^2\sigma$ or $r\dot{\gamma}/D_s$ where $D_s = k_BT/(6\pi\mu r)$ is the sphere diffussivity according to the Stokes-Einstein relationship (where k_B is the Boltzmann constant and T is the absolute temperature.) (D and F). Microspheres of three different radii and/or media of four different viscosities were used (indicated). The data were recorded at 250 frames per second. Reproduced from Ref. [9].

We have seen that plotting $p_{ad}/(m_rm_lr)$ vs. $r\dot{\gamma}$ (Fig. 3, C and E) and vs. $r\dot{\gamma}/D_s$ (Fig. 3, D and F) respectively aligned the initial portions of the ascending phase and the final portions of the descending phase of different curves corresponding to different parameters. This observation suggests that combining $r\dot{\gamma}$ and $r\dot{\gamma}/D_s$ will provide an abscissa variable to align both phases provided that the ordinate is also rescaled. To construct such a combined variable, we note that, when plotted against D_s, the $D_s/r\dot{\gamma}$ value where $p_{ad}/(m_rm_lr)$ achieved maximum (referred to as optimal $D_s/r\dot{\gamma}$) can be fitted with two lines, $(D_s/r\dot{\gamma})_{opt} = A(D_s + C_1)$, one for microspheres (with C_1 = 0.09 μm²/s) and the other for neutrophils (with 0.05 μm²/s), where A equals 1.45 μm/s for both data sets (Fig. 4B). Multiplying the fitting equation by $r\dot{\gamma}/D_s$ resulted in $(r\dot{\gamma}/D_s)(D_s/r\dot{\gamma})_{opt} = A\,r\dot{\gamma}\,(1 + C_1/D_s)$. The right-hand side is a linear combination of $r\dot{\gamma}$ and $r\dot{\gamma}/D_s$. On the left-hand side, $r\dot{\gamma}/D_s$ is scaled by its optimal value, which should align the locations where $p_{ad}/(m_rm_lr)$ peaks when plotted against the combined variable.

2.4 *Enhancing tethering by molecular diffusion*

As the medium viscosity decreased, both the $r\dot{\gamma}$ value where $p_{ad}/(m_rm_lr)$ achieved maximum and the maximum $p_{ad}/(m_rm_lr)$ value increased (Fig. 3, C and E). However, the maximum $p_{ad}/(m_rm_lr)$ value was insensitive to the sphere radius (Fig. 3B) and did not correlate with the sphere diffusivity. We suggested that the governing factor is molecular diffusivity, $D_m \equiv k_BT/(6\pi\mu l)$ calculated from the Stokes-Einstein

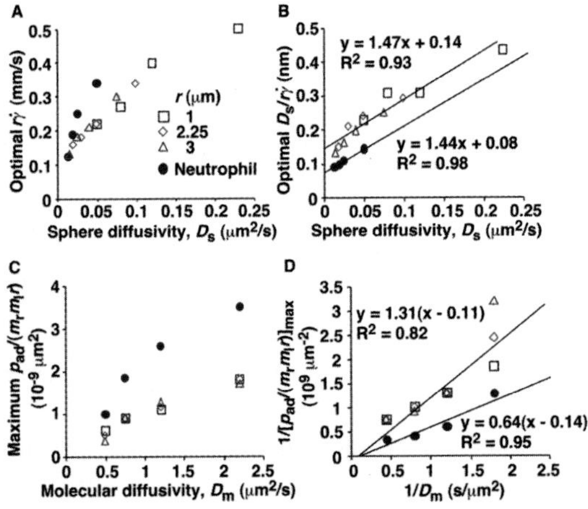

Figure 4. Analysis of optimal values of tether rate curves. A, Peak locations of the $p_{ad}/(m_r m_l r)$ vs. $r\dot{\gamma}$ curves (optimal $r\dot{\gamma}$) were plotted against the sphere diffusivity D_s. B, Peak locations of the $p_{ad}/(m_r m_l r)$ vs. $D_s/ r\dot{\gamma}$ curves (optimal $D_s/ r\dot{\gamma}$) were plotted against the sphere diffusivity D_s. C, Maximum $p_{ad}/(m_r m_l r)$ values were plotted against the molecular diffusivity D_m. D, Reciprocal of maximum $p_{ad}/(m_r m_l r)$ values were plotted against reciprocal of the molecular diffusivity. Positive correlations were evident in all plots for both microspheres (open symbols) and neutrophils (solid circles). A straight line was fit to each set of the data for microspheres or neutrophils in B and D. The best-fit equations are indicated along with the R^2 values. Reproduced from Ref. [9].

relationship, where l is a characteristic length in the molecular scale, e.g., l = 100 nm for an order-of-magnitude estimate. We plotted the maximum $p_{ad}/(m_r m_l r)$ value against D_m. These plots were nearly linear for both microspheres and neutrophils regardless of the sphere radius (Fig. 4C), suggesting that molecular diffusion is another limiting factor for tethering. Scaling $p_{ad}/(m_r m_l r)$ by its maximum value determined by molecular diffusivity, as is done using Fig. 4D, should collapse the ordinate of tethering curves (Fig. 5, C and D), just as scaling ($r\dot{\gamma}/D_s$) by its optimal value collapses the abscissa range of these curves (Fig. 5, A and B).

Thus, dimensional analysis identifies three distinct transport modes – sliding of

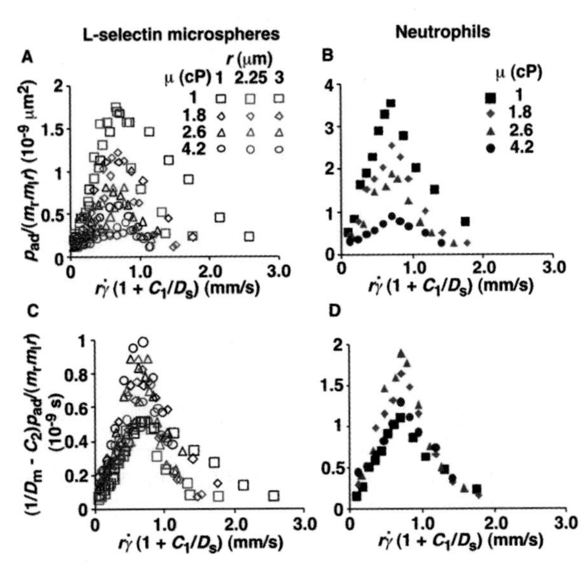

Figure 5. Collapse of multiple data curves by proper scaling of the contributions by three transport mechanisms. When the normalized adhesion probabilities per distance for microspheres (A) and neutrophils (B), $p_{ad}/(m_r m_l r)$, were plotted vs. $r\dot{\gamma}$ $(1 + C_1/D_s)$, a variable that combines sphere transport mechanisms for both relative sliding and Brownian motion, the ranges of all curves were aligned. When the $p_{ad}/(m_r m_l r)$ values were further multiplied by $(1/D_m - C_2)$ to obtain a variable that combines molecular diffusion and molecular docking, all 12 microsphere curves collapsed into a single curve (C). Similarly, all 4 neutrophil curves collapsed into a single curve (D). Reproduced from Ref. [9].

the cell bottom on the vascular surface, cell Brownian motion, and molecular diffusion – that enhance cell tethering. When transport is sufficiently fast, bond formation becomes reaction-limited. Further increase in transport shortens the contact durations, which decreases cell tethering. Taking together, our analysis fully explains the biphasic tether rate vs. flow curve (Fig. 1A).

3 Catch Bonds Govern Flow-enhanced Cell Rolling

3.1 *Conceptual scheme of rolling process*

Let us now turn to another aspect of flow-enhanced adhesion: rolling velocity decreases with increasing flow until an optimal wall shear stress is reached, after which cells roll more rapidly as flow increases (Fig. 1C). Fig. 6 illustrates the physical parameters of cell rolling under flow. Shear stress σ, the product of shear rate $\dot{\gamma}$ and viscosity μ, applies a resultant force F_s and a torque T_s to the adherent sphere (or cell) (Fig. 6A), both of which increase when the rolling sphere slows and reach their maximums when the sphere stops. F_s and T_s are balanced by tensile forces (F_t) applied to the adhesive bonds at the trailing edge of a tethered sphere, and by compressive forces at the sphere bottom (F_c). Shear stress-dependent F_c could enlarge the cell-surface contact area thus facilitating the binding for the deformable cell but not for the fixed cell or rigid microsphere. F_t reaches maximum when the sphere stops, $[F_t]_{max} = 13.2r^2\sigma$ [6, 11].

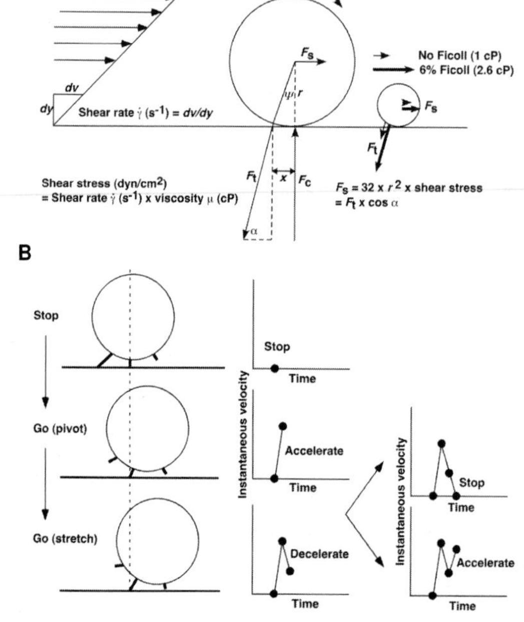

Figure 6. Parameters of cell rolling under flow. A, Balance of forces acting on a rolling cell. Shear stress applies a resultant force F_s and a torque T_s to the cell, which are balanced by a tether force F_t on the receptor-ligand bonds at the trailing edge of a tethered cell and by a compressive force F_c at the sphere bottom. Elevating the viscosity by addition of 6% Ficoll increases shear stress by 2.6-fold and increase F_t on the sphere of the same size, as illustrated by the comparative lengths of the thin and thick vectors for F_t on the small sphere on the right. At the same shear stress, F_t is 9-fold greater for a sphere of 3-μm radius then for a sphere of 1-μm radius, as illustrated by the comparative lengths of the thin vectors for F_t on the large and small spheres. The conversion of wall shear stress into F_t is described in [10]. B, Decomposition of a cyclic rolling step. See text for a detailed description. Reproduced from Ref. [11].

Cell rolling is characterized by formation of new bonds at the front edge and dissociation of preexisting bonds at the rear edge. Fig. 6B illustrates a minimal model for a cyclic rolling step that alternates between one and two bonds. The cell stops when the force on the rear bond reaches $[F_t]_{max}$, which allows formation of another bond a step distance ahead. Once the rear bond ruptures, the cell rolls forward with increasing velocity, i.e., acceleration. After pivoting over the front bond, which is being translated to the trailing edge, the cell decelerates by a tensile force that stretches the bond. If this bond is sufficiently strong to withstand the force $[F_t]_{max}$, the cell stops again. If the bond ruptures prematurely, the cell accelerates again. This cycle repeats as the cell rolls continuously.

3.2 Rolling velocity scales with tether force

Again, we used dimensional analysis to identify the governing parameter that underlies the mechanism(s) for flow-enhanced cell rolling. Rolling velocities of fixed or unfixed cells or microspheres of different radii r in media of different viscosities μ were measured at different wall shear rates $\dot{\gamma}$. When plotted against $\dot{\gamma}$, the biphasic rolling velocity curves shifted horizontally with r and/or μ for both microspheres (Fig. 7A) and cells (Fig. 7D). When the abscissa was rescaled by multiplying $\dot{\gamma}$ by μ to plot the rolling velocity against the wall shear stress $\sigma = \mu\dot{\gamma}$, curves for different μ aligned for microspheres of the same r (but shifted with r) (Fig. 7B) and for neutrophils prepared the same way (fixed or unfixed) (Fig. 7E). When the abscissa was rescaled again by multiplying σ by $13.2r^2$ to plot the rolling velocity against the maximum tether force $[F_t]_{max} = 13.2r^2\sigma$, all 4 microsphere curves (Fig. 7C) collapsed, indicating that $[F_t]_{max}$ is the governing parameter.

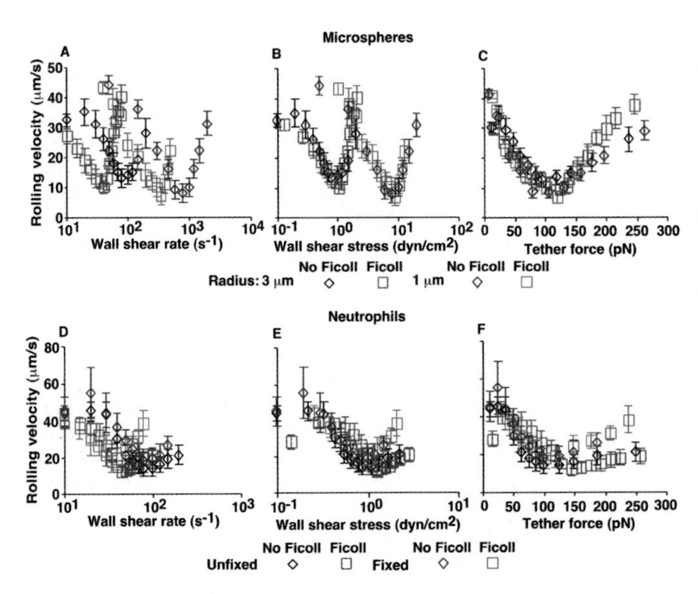

Figure 7. Mean velocities of L-selectin microspheres (A-C) or of unfixed or fixed neutrophils (D-F) rolling on PSGL-1 in the absence or presence of 6% Ficoll (which increased the viscosity by 2.6 fold) were plotted against wall shear rate (A and D), wall shear stress (B and E), and tether force (C and F). Reproduced from Ref. [11].

The normal force F_c at sphere bottom also increases with the tether force. The compression by this force may enlarge the contact area, which enables more bonds to form. In addition, force acting on adhesive bonds can extrude membrane tethers from live cells [12, 13]. Both mechanisms can slow rolling velocity and both scale with tether force. We tested these hypothetical mechanisms by altering the cell deformability using fixation and using rigid microspheres. When plotted against wall shear stress or tether force, the curves of unfixed and fixed neutrophils showed only moderate differences beyond the optimal shear level; both sets of curves were comparable to the microsphere curves, especially in the regime below the optimal tether force (Fig. 7, C and F). These results suggest that enlargement of contact area and/or extrusion of membrane tethers, if occurred, play only a minor role in the flow regime tested.

3.3 Off-rate curves and rolling velocity curves correlate and scale similarly

Discussion on the rolling step model (Fig. 6B) suggests the importance of off-rate for dissociation of receptor-ligand bonds in controlling the rolling velocity. Only after the rearmost bond dissociates will the cell roll a step forward. Thus, the slower the off-rate is, the slower the rolling velocity is. We previously demonstrated that selectin-ligand interactions exhibit catch-slip transitional bonds where increasing force initially decreases off-rate and then increases off-rate, which may explain the biphasic rolling velocity vs. force curves [14, 15]. To obtain further support of this hypothesis, we used dimensional analysis again.

Using dimensional analysis, we first confirmed that tether force, not wall shear rate or wall shear stress, is the governing parameter for receptor-ligand dissociation. To do that we measured transient tether lifetimes of fixed and unfixed neutrophils and microspheres of different radii in media of different viscosities flowing in

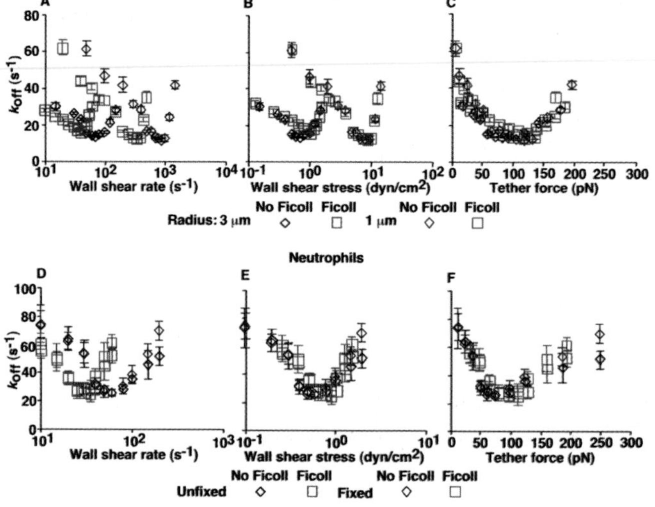

Figure 8. Off-rates (k_{off}) derived from lifetimes of transient tethers of L-selectin-bearing microspheres of 3-μm and 1-μm radii (A-C) or of unfixed or fixed neutrophils (D-F) to low density of sPSGL-1 (<10 sites/μm^2) in the absence or presence of 6% Ficoll (which increased the medium viscosity by 2.6 fold) were plotted against wall shear rate (A and D), wall shear stress (B and E), and tether force (C and F). Reproduced from Ref. [11].

different wall shear rates. The off-rates of single L-selectin-PSGL-1 bonds were determined from lifetime measurements and plotted against wall shear rate (Fig. 8, A and D), wall shear stress (Fig. 8, B and E) and tether force (Fig. 8, C and F). As expected, off-rate curves display catch-slip transitional bonds and align only when plotted against the tether force. The similar shape and common scaling law of both the rolling velocity and the off-rate curves thus support a causal relationship between catch bonds and flow-enhanced rolling.

3.4 Off-rate curves and curves of multiple rolling regularity metrics correlate and scale similarly

Further support for the proposed causal relationship between catch bonds and flow-enhanced rolling was obtained from detailed analysis of the stop-and-go rolling step cycles illustrated schematically in Fig. 6B. Fig. 9 shows the time courses of instantaneous velocities of representative microspheres of 3-μm radius bearing L-selectin freely flowing over a surface coated with a nonadhesive protein human serum albumin (HSA) (first panel) or continuously rolling on a surface coated with the adhesive ligand PSGL-1 (other panels) in media without Ficoll in a range of wall shear rates measured with 4 ms temporal resolution. The instantaneous velocities of microspheres rolling on PSGL-1 alternated between zero and nonzero values, whereas the velocities of microspheres freely flowing over HSA fluctuated much less and never dropped to zero. The mean of the free-flowing velocities increased with increasing wall shear rate, whereas their fluctuations remained at similarly

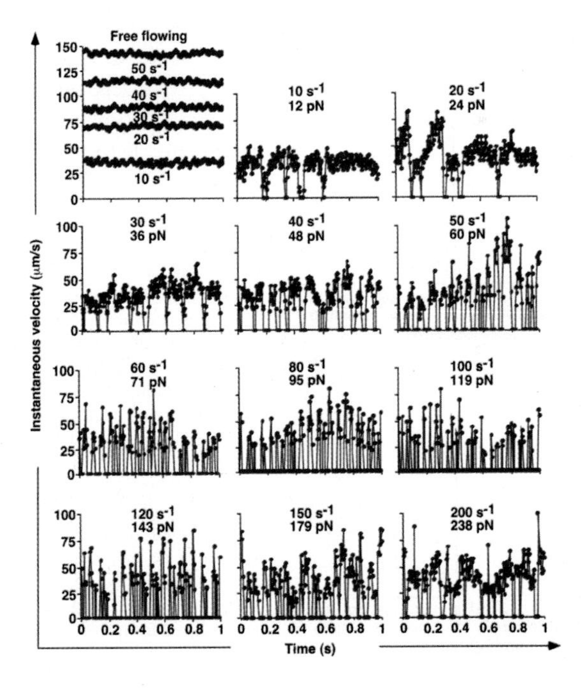

Figure 9. Changing features of instantaneous velocities of L-selectin-bearing mocrospheres freely flowing over HSA (1st panel) or continuously rolling on PSGL-1 (other panels) at indicated wall shear rates (s^{-1}) and corresponding tether forces (pN). The microsphere radius was 3 μm and the media did not contain Ficoll. The data were recorded at 4-ms temporal resolution. Reproduced from Ref. [11].

small levels, which could be attributed to Brownian motion. By contrast, the mean of the rolling velocities initially decreased and then increased with increasing shear because motions were interrupted by stops whose frequency and duration initially increased and then decreased. As the shear rate increased toward the optimal value, the rolling velocity fluctuations intensified, the acceleration-deceleration cycles became more regular, and more decelerations converted into full stops with zero velocity, which became longer, indicating stronger/longer-lived L-selectin-PSGL-1 bonds. As flow increased above the optimal value, these trends were reversed. The rolling velocity fluctuations became less intensive and less regular, and fewer decelerations converted into full stops, which became shorter, indicating weaker/shorter-lived L-selectin-PSGL-1 bonds.

The above qualitative observations were quantified using an acceleration threshold to segregate the velocity curves into periods of stops and go's, thereby allowing statistical analysis of a variety of metrics for rolling regularity and stability [11]. These include mean stop and go times, stop and go frequencies, fractional times spent in the stop or go phases, fractions of steps with stops, and mean go distance. To perform dimensional analysis, nearly a thousand stop and go events were collected for each condition from 10-15 microspheres of 1-μm or 3-μm radii or fixed or unfixed neutrophils that rolled continuously for 1 s in media with or without Ficoll in a range of shear rates. As exemplified by stop frequency (Fig. 10) and mean stop time (Fig. 11), plotting these rolling parameters vs. wall shear rate, wall shear stress and tether force revealed that all of the curves aligned with the tether force but not with wall shear rate or wall shear stress. In addition, all curves were biphasic, showing increased rolling regularity and stability as force increased toward the optimal level but reversed the trends beyond the optimal force.

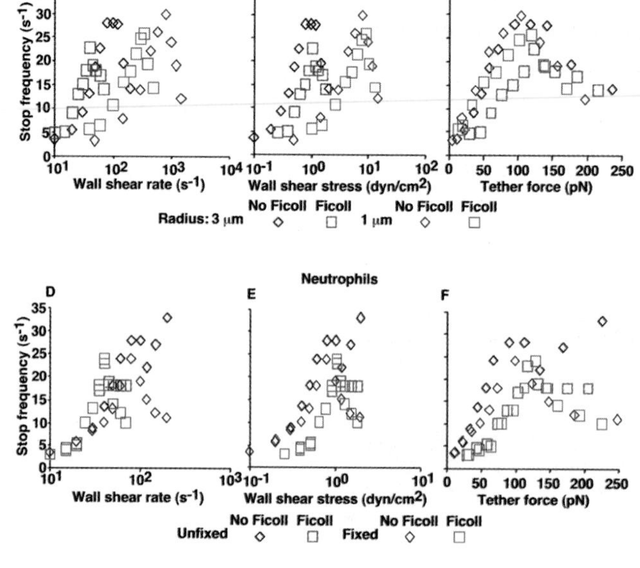

Figure 10. Tether force governs stop frequency below and above the flow optimum. The stop frequency (number of stops in a 1-s time course of instantaneous velocities as those shown in Fig. 9) of L-selectin-bearing microspheres of 3- or 1-μm radii (A-C) and fixed or unfixed neutrophils (D-F) rolling on sPSGL-1 in the absence or presnce of 6% Ficoll (corresponding to viscosity of 1 or 2.6 cP) were plotted against wall shear rate (A and D), wall shear stress (B and E), and tether force (C and F). Reproduced from Ref. [11].

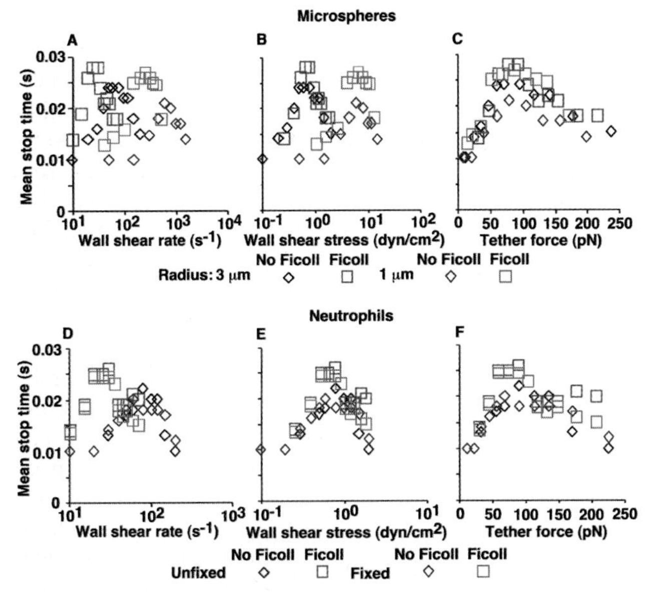

Figure 11. Tether force governs mean stop time below and above the flow optimum. The mean stop times (durations of stops in the time courses of instantaneous velocities as those shown in Fig. 9) of L-selectin-bearing microspheres of 3- or 1-µm radii (A-C) and fixed or unfixed neutrophils (D-F) rolling on sPSGL-1 in the absence or presence of 6% Ficoll (corresponding to viscosity of 1 or 2.6 cP) were plotted against wall shear rate (A and D), wall shear stress (B and E), and tether force (C and F). Reproduced from Ref. [11].

Thus, in all areas examined, strong correlations were observed between off-rate of L-selectin-PSGL-1 dissociation and multiple metrics of rolling regularity and rolling stability. These include the optimal tether force level where parameters achieved extreme values (maximum or minimum), the trends of changes below and above the optimal tether force, and even the shapes of the curves. More importantly, the changes in the rolling regularity and stability can be understood intuitively and explained easily by the changes in the off-rate. These observations and analyses unambiguously demonstrated catch bonds as the mechanism for flow-enhanced rolling. It is this unusual property of selectin-ligand interaction that converts the weak and short-lived bonds into strong and longer-lived bonds, which causes the counterintuitive decrease in the rolling velocity and increase in the rolling regularity and stability as shear rises from the threshold to an optimal value.

4 Discussion and Conclusion

Since its discovery, flow-enhanced cell adhesion has long intrigued researchers because it appears paradoxical. Using dimensional analysis, we have identified two mechanisms of flow-enhanced leukocyte adhesion mediated by selectin-ligand interactions: transport-dependent acceleration of bond formation and force-dependent deceleration of bond dissociation. Dimensional analysis is usually used in mechanics for reduction of simpler relationships among smaller number of (dimensionless) variables than those in the (dimensional) governing equations. Here, we used it for induction of relevant parameters underlying the dominant mechanisms.

This was done in a manner resembling hypothesis testing based on experimental data rather than derivation based on first principles. The mechanisms of selectin-mediated flow-enhanced leukocyte adhesion identified by dimensional analysis have been further confirmed by mutagenesis studies where point mutation was used to increase the rotational diffusivity and argument the catch bonds of L-selectin [16]. The degrees of flow enhancements for cell tethering and for cell rolling have been observed to increase correspondingly [16]. The experimental strategies and analysis methods described here and in Ref. [16] may be applicable to other biological systems that exhibit flow-enhanced adhesion, for instance, GPIb-expressing platelet adhesion to VWF and FimH receptor-expressing bacterium adhesion to mannosylated ligands, which may be governed by mechanisms similar to those identified here. Studies are under way to test these predictions. Future studies will test these predictions before long.

Acknowledgments

We gratefully acknowledge the contributions of our co-workers who produced the original data in Refs. [9, 11]. This work was supported by National Institutes of Health grants AI44902 (CZ) and HL65631 (RPM)

References

1. Finger, E.B., Puri, K.D., Alon, R., Lawrence, M.B., von Andrian, U.H., Springer, T.A., 1996. Adhesion through L-selectin requires a threshold hydrodynamic shear, Nature 379, 266-269.
2. Lawrence, M.B., Kansas, G.S., Kunkel, E.J., Ley, K., 1997. Threshold levels of fluid shear promote leukocyte adhesion through selectins (CD62L,P,E), J. Cell Biol. 136, 717-727.
3. Savage, B., Saldivar, E., Ruggeri, Z.M., 1996. Initiation of platelet adhesion by arrest onto fibrinogen or translocation on von Willebrand factor, Cell 84, 289-297.
4. Doggett, T.A., Girdhar, G., Lawshe, A., Miller, J.L., Laurenzi, I.J., Diamond, S.L., Diacovo, T.G., 2003. Alterations in the intrinsic properties of the GPIbalpha-VWF tether bond define the kinetics of the platelet-type von Willebrand disease mutation, Gly233Val, Blood 102, 152-160.
5. Thomas, W.E., Trintchina, E., Forero, M., Vogel, V., Sokurenko, E.V., 2002. Bacterial adhesion to target cells enhanced by shear force, Cell 109, 913-923.
6. Goldman, A.J., Cox, R.G., Brenner, H., 1967. Slow viscous motion of a sphere parallel to a plane wall. II. Couette flow, Chem. Eng. Sci. 22, 653-660.
7. Happel, J., Brenner, H., 1991. Low reynolds number hydrodynamics. Kluwer Academic Publishers, Dordrecht, The Netherlands.

8. Mege, J.L., Capo, C., Benoliel, A.M., Bongrand, P., 1986. Determination of binding strength and kinetics of binding initiation. A model study made on the adhesive properties of P388D1 macrophage-like cells, Cell Biophys. 8, 141-160.
9. Yago, T., Zarnitsyna, V.I., Klopocki, A.G., McEver, R.P., Zhu, C., 2007. Transport governs flow-enhanced cell tethering through L-selectin at threshold shear, Biophys. J. 92, 330-342.
10. Yago, T., Leppänen, A., Qiu, Hy., Marcus, W.D., Nollert, M.U., Zhu, C., Cummings, R.D., McEver, R.P., 2002. Distinct molecular and cellular contributions to stabilizing selectin-mediated rolling under flow, J. Cell Biol. 158, 787-799.
11. Yago, T., Wu, J., Wey, C.D., Klopocki, A. G., Zhu, C., McEver, R.P., 2004. Catch bonds govern adhesion through L-selectin at threshold shear, J. Cell Biol. 166, 913-923.
12. Ramachandran, V., Williams, M., Yago, T., Schmidtke, D.W., McEver, R.P., 2004. Dynamic alterations of membrane tethers stabilize leukocyte rolling on P-selectin, Proc. Natl. Acad. Sci. USA 101, 13519-13524.
13. Schmidtke, D.W., Diamond, S.L., 2000. Direct observation of membrane tethers formed during neutrophil attachment to platelets or P-selectin under physiological flow, J. Cell Biol. 149, 719-730.
14. Marshall, B.T., Long, M., Piper, J.W., Yago, T., McEver, R.P., Zhu, C., 2003. Direct observation of catch bonds involving cell-adhesion molecules. Nature 423, 190-193.
15. Sarangapani, K.K., Yago, T., Klopocki, A.G., Lawrence, M.B., Fieger, C.B., Rosen, S.D., McEver, R.P., Zhu, C., 2004. Low force decelerates L-selectin dissociation from P-selectin glycoprotein ligand-1 and endoglycan, J. Biol. Chem. 279, 2291-2298.
16. Lou, J., Yago, T., Klopocki, A.G., Mehta, P., Chen, W., Zarnitsyna, V.I., Bovin, N.V., Zhu, C., McEver, R.P., 2006. Flow-enhanced adhesion regulated by a selectin interdomain hinge, J. Cell Biol. 174, 1107-1117.

A SLIDING-REBINDING MECHANISM FOR CATCH BONDS

J. LOU AND C. ZHU

Coulter Department of Biomedical Engineering, Woodruff School of Mechanical Engineering, and Institute for Bioengineering and Bioscience, Georgia Institute of Technology, Atlanta, GA 30332, USA
E-mail: cheng.zhu@bme.gatech.edu

T. YAGO AND R. P. MCEVER

Cardiovascular Biology Research Program, Oklahoma Medical Research Foundation, and Department of Biochemistry and Molecular Biology, University of Oklahoma Health Sciences Center, Oklahoma City, OK 73104, USA

The dissociation kinetics of biomolecular interactions have long been believed to behave as slip bonds such that lifetimes of a molecular complex (or a "molecular bond") decrease with increasing externally applied force. This behavior can be explained with intuitive physical models and has been observed in many experiments. Recently, the opposite behavior, termed catch bonds based on their theoretical possibility, has been demonstrated experimentally, where bond lifetimes increase with force. This discovery has sparked great interest in physically modeling such counterintuitive phenomena. Here we summarize our recent work with selectin crystal structures, molecular dynamics simulations, Monte Carlo modeling, site-directed mutagenesis, single-molecule force and kinetics experiments, and flow chamber adhesion studies that support a sliding-rebinding mechanism for catch bonds. In the model, "catch" results from forced opening of an interdomain hinge that tilts the binding interface to allow two sides of the contact to slide against each other. Sliding promotes formation of new interactions and even rebinding to the original state, thereby slowing dissociation and prolonging bond lifetimes. This model provides a possible explanation for how external forces allosterically modulate atomic-level noncovalent interactions at the binding interface to regulate the dissociation pathways.

1 Introduction

Interactions among biomolecules generally occur as rapid transitions from bound and unbound states; hence they are described by reaction kinetics. The kinetics of biomolecular interactions may be regulated by mechanical forces. External force may be particularly important in the interaction kinetics of those molecules whose functions include a mechanical role, for example, cell adhesion molecules that provide mechanical linkage between cells or between cells and extracellular matrices. How force affects the kinetics of biomolecular interactions is therefore a fundamental question in biology. The dissociation of biomolecular interactions is usually described using the framework of chemical reaction kinetics theory. The force regulation of biochemical reaction is generally treated mathematically by the force-dependence of kinetic rates. In 1978 George Bell proposed the first model for force-dependence of off-rate of receptor-ligand dissociation, where off-rate is expressed as an exponentially increasing function of the external force [1]. Ten years later, Dembo et al. theorized that off-rate could be expressed as an exponentially decreasing function of the external force [2]. Biological bonds were categorized based on their force-dependence of off-rates: catch bonds if off-rate increases with force, slip bonds if off-rate decreases with force, or ideal bonds if off-rate does not change with force [2]. In 1995 the first experimental demonstration of slip bonds was performed for P-selectin interacting with P-selectin glycoprotein ligand 1 (PSGL-1), with the results analyzed using the Bell model [3]. Since then, many more experiments showing slip bond behavior with other molecular systems have been published. In 1997 Evans and Ritchie extended Kramers' theory for reaction kinetics [4] to provide a solid foundation for the Bell equation to describe slip bonds [5].

Not until 2003 were catch bonds demonstrated experimentally, again for the P-selectin-PSGL-1 interaction [6]. Force elicited catch bonds until an optimal force was achieved where bond lifetimes were longest; further increases in force resulted in a transition to slip bonds. Catch-slip transitional bonds have also be observed for L-selectin-ligand interactions [7, 8] and actin-myosin interactions [9] and suggested for other systems [10]. The observations of the intriguing catch bonds have sparked theoretical interest. Several models have been proposed to explain this counterintuitive phenomenon (see Ref. [11] for a review). Most of these models assume receptor-ligand dissociation as diffusive escape of a particle from an energy well and also assume two-pathway dissociation.

At the atomic level, a biomolecular interaction involves many noncovalent interactions, such as electrostatic interactions, van der Waals interactions, hydrogen bonds, and dipole-dipole interactions. The cooperation of the individual interactions governs the kinetics of the whole "biological bond." To truly understand biomolecular interactions in general and catch bonds in particular, we should look at the binding interface in detail, including the atomic-level structure and its change under external forces.

In this review we summarize our sliding-rebinding model for catch bonds [8, 12]. It is based on studies of crystal structures of selectins and selectin-ligand complexes. It has been supported by steered molecular dynamics (MD) simulations and mutagenesis studies. We will also discuss the relationship between the sliding-rebinding mechanism and other catch bond models.

2 Structures of Selectins and Selectin-Ligand Complexes

Six selectin (lectin and EGF domains) crystal structures have been solved: unliganded E-selectin [13], P-selectin [14], and L-selectin (P. Mehta, V. Oganesyan, S. Terzyan, T. Mather, and R.P. McEver, personal communication), E-selectin and P-selectin respectively liganded with the glycan determinant sialyl Lewis x (sLe^x)

[14], and P-selectin liganded with an N-terminal glycosulfopeptide fragment of PSGL-1, which comprises sLe^x on an O-glycan arrayed in a stereochemically precise manner near sulfated tyrosines and other peptide components [14]. Consistent with the highly conserved sequences (>60% identities and >70% positives), the three unliganded selectin lectin-EGF structures are highly similar, with very small backbone root mean square distance (RMSD) between the lectin-EGF domains (residues 1-155) of different selectins (0.73, 1.19, and 1.39Å between P- and E-selectin, between E- and L-selectin, and between P- and L-selectin, respectively). The binding of sLe^x did not change the lectin-EGF conformation for P- or E-selectin. The RMSD between the two lectin-EGF structures with and without sLe^x binding are only 0.33 Å (backbone) and 0.43 Å (all heavy atoms) for P-selectin and 0.24 Å (backbone) and 0.60 Å (all heavy atoms) for E-selectin. However, the lectin and EGF domains of P-selectin show significant conformational changes when bound to PSGL-1. The RMSD between the unliganded and liganded structures are 1.77 Å for the lectin domain and 17 Å for the EGF domain when the two lectin domains are aligned. The differences in the two EGF domain conformations are shown in Fig. 1A,

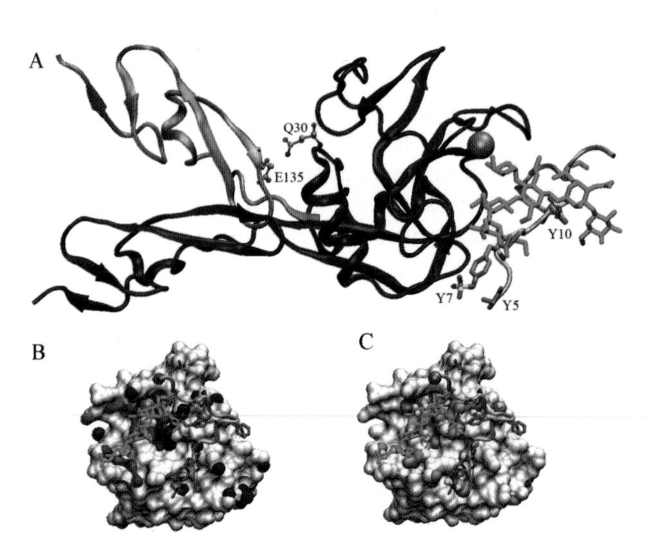

Figure 1. *A*, Co-crystal structures of P-selectin complexed with ligands. The open-angle structure of P-selectin (blue, lectin and EGF domains) bound to PSGL-1 is shown by ribbon representation (PDB code 1G1S). The closed-angle structure of P-selectin (mauve, only EGF domain shown) bound to sLe^x (not shown) is superimposed (PDB code 1G1R). The two lectin domains have been aligned and only the one that was complexed with PSGL-1 is shown. A hydrogen bond between Gln30 and Glu135 (shown by sticks and balls) was observed in the closed-angle but not the open-angle structure. A calcium ion is shown as a golden sphere. The PSGL-1 peptide is shown in pink. Three sulfated tyrosines (indicated) are highlighted as cyan sticks with yellow (sulphur) and red (oxygen) ends on the side chains of Tyr7 and Tyr10. The side chain of sulfated Tyr5 is not shown as it was missing from the crystal structure. Glycan is shown in green. *B*, Top view of the binding surface. The lectin domain is shown by electrostatic potential surface representation. The nitrogen atoms of the positively charged residues are colored in blue and the oxygen atoms of the hydrogen bond donor/acceptor residues are colored in red. PSGL-1 glycan is shown in green and peptide is shown in pink with the side chains of the negatively charged residues shown in red and those of the hydrophobic residues shown in purple. The calcium ion is shown as a large golden sphere. *C*, Noncovalent interactions between the lectin domain atoms (cyan) and PSGL-1 atoms (yellow). The C_α atom of PSGL-1 Pro18 is marked by an ice-blue sphere. Reproduced from Ref. [12]

where the lectin-EGF domains of unliganded P-selectin and P-selectin in complex with PSGL-1 (only the EGF domain is shown) are aligned together. Unliganded P-selectin displays a more "closed" angle between the lectin and EGF domains, which is stabilized by a hydrogen bond between residues Q30 and E135. In complex with PSGL-1, the lectin-EGF hinge assumes a more "open" conformation.

P-selectin binds PSGL-1 through a broad and shallow interface that contains many atomic-level contacts [14-16]. The sugar segment of PSGL-1 docks to a more acidic surface, where many oxygen atoms on the sugar ring can potentially form hydrogen bonds with hydrogen bond acceptor/donor residues on the lectin. The peptide segment is highly negatively charged and docks to a more basic surface, which favors interactions between the negative residues on PSGL-1 and positive residues on lectin surface (Fig. 1B). The interaction network is shown in Fig. 1C. The strongest contact is the metal ion coordination of O3 and O4 atoms from the fucose residue. The side chains of lectin residues Glu80 (with O4 atom), Asn84 (with O4 atom), Glu88 (with O2 atom) and Glu107 (with O3 atom) also form hydrogen bonds with the fucose, which stabilizes the Ca^{2+} coordination. The galactose residue forms hydrogen bonds with lectin Glu92 and Tyr94, and the sialic acid residue (NeuAc) forms hydrogen bonds with lectin residues Tyr48 and Ser99. The peptide portion of PSGL-1 interacts with the lectin domain mainly through its sulfated tyrosines. The sulfated Tyr7 interacts with the backbone oxygen atoms of S46 and S47 and also with the side chain of His114 of lectin. Sulfated Tyr10 interacts with lectin Arg85 which also forms a hydrogen bond with the backbone oxygen atom of the PSGL-1 Pro14. The backbone oxygen atom of sulfated Tyr7 forms a hydrogen bond with the backbone nitrogen atom of lectin Lys112 whose side chain also forms a hydrogen bond with PSGL-1 Glu6. The hydrophobic side chains of PSGL-1 Pro14, Leu13 and Leu8 are packed against lectin domain residues His108, Leu110 and Lys111. Thus, the stable bound state is a result of many atomic-level, noncovalent interactions.

3 MD Simulations of Selectins and Selectin-Ligand Complexes

Two types of MD simulations were performed using NAMD package [17] with the CHARMM22 all atom force field [18]. One type is free dynamics simulations of the lectin-EGF domains in either closed or open conformation using either P- or L-selectin. The other type is forced unbinding (steered MD or SMD) of P-selectin-sLe^x and P-selectin-PSGL-1 complexes.

3.1 Free dynamics simulations of selectin lectin-EGF domains

The free dynamics simulations were focused on the stability of the "open" and "closed" conformations as well as possible transitions between the two conformations. The simulations were performed on the closed P- and L-selectin as well as the open P-selectin. For both open and closed P-selectin simulations, we found that the structure fluctuated around the original positions and the conformations were well maintained, suggesting that the two conformations are stable (Fig. 2A). Most interestingly, transition to the open conformation was observed in the simulations of closed L-selectin, as shown in Fig. 2B, where the RMSD of the EGF domain between the simulated structures and either the closed or open structure are plotted vs. simulation time. These data suggest that simultanous transitions between the open and closed conformation are possible.

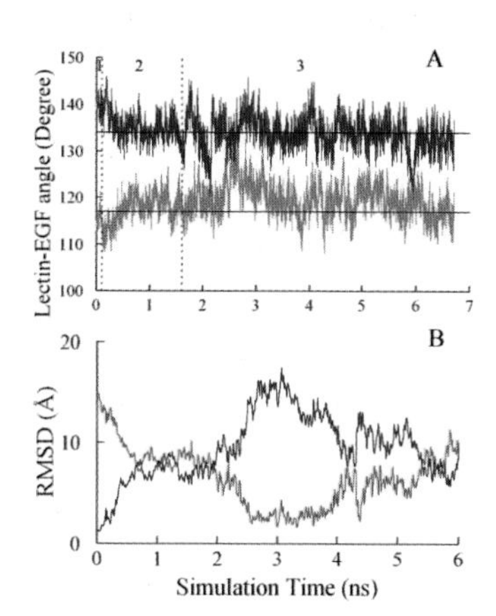

Figure 2. *A*, MD-simulated time courses of lectin-EGF interdomain angles. Simulations started from the closed-angle unliganded P-selectin structure (mauve) or the open-angle liganded P-selectin structure (blue) after deleting the PSGL-1. The two horizontal lines indicate the corresponding angles observed in the crystal structures. Separated by dotted vertical lines, regions 1, 2 and 3 mark, respectively, times for heat-up, equilibration, and free dynamics simulations. Reproduced from Ref. [12]. *B*, RMSD between the corresponding backbone atoms from residues 121–156 (the EGF domain) of the simulated L-selectin structure and the crystal structures of closed-angle L-selectin (blue curve) or open-angle P-selectin (red curve) as a function of simulation time. The lectin domains were aligned by minimizing the RMSD between the backbone atoms from residues 1–120. Reproduced from Ref. [8].

3.2 SMD simulations of unbinding of selectin-ligand complexes

To study selectin-ligand dissociation under force, we applied an external force in MD to simulate unbinding of P-selectin (lectin-EGF domains or lectin domain alone) from ligand (PSGL-1 or sLe^x). Simulations with sLe^x began with the closed lectin-EGF structure whereas those with PSGL-1 began with the open lectin-EGF structure. In both cases, increasing force gradually opened the lectin-EGF angle (Fig. 3A). Simulations also revealed that the lectin-EGF angle could assume several conformations in addition to the two observed in the crystal structures (Fig. 3A).

In simulations with the lectin-EGF domains, it was usually observed that the EGF domain unfolded before ligand dissociation from the lectin domain. To avoid possible adverse effects caused by unfolding, most simulations used the lectin domain only. In one simulation with both the lectin and EGF domains, the EGF domain did not unfold before ligand dissociation. Fig. 3B shows the force-time cource of this simulation. The force peaked at a much higher level than those observed in experiments, probably because unbinding was accelerated to occur in a time scale at least 6 orders of magnitudes shorter than the laboratory time scale due to limited computational resources. The preexisting atomic-level interactions began to break after 1 ns. By ~2.7 ns, the last preexisting interaction – the Ca^{2+} coordination with the fucose residue – ruptured, which initiated the dissociation of the interface. New interactions were observed to form before and after this initial dissociation. The interactions formed after the initial dissociation seem most interesting and important, as they kept PSGL-1 from being completely dissociated from P-selectin for 1 ns at a smaller force of ~200 pN.

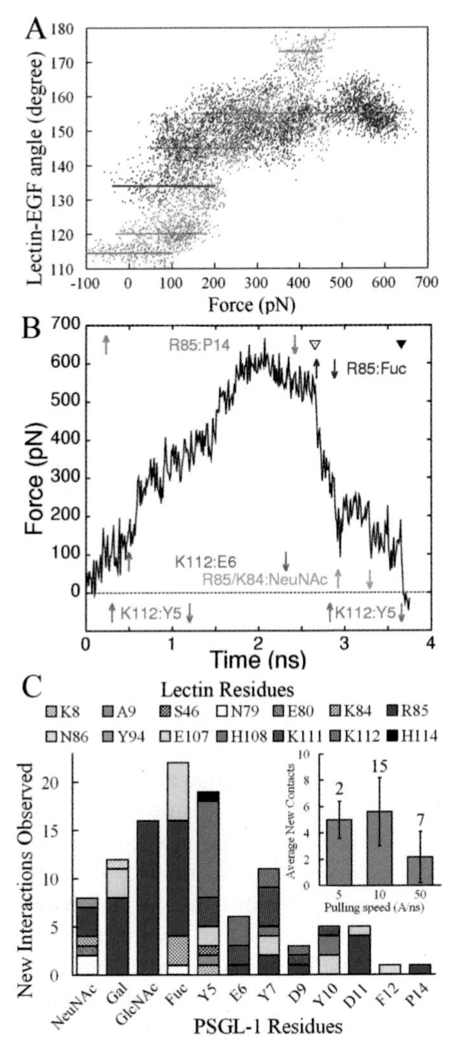

Figure 3. *A*, Interdomain angle vs. pulling force plots. Data from two SMD simulations are overlaid: a 3.6-ns simulation starting from the closed-angle structure of P-selectin liganded with sLe^x (mauve dots) and a 3.75-ns simulation starting from the open-angle structure of P-selectin liganded with PSGL-1 (blue dots). The C_α atom of Gly147 of the EGF domain was constrained, and a force was applied through a spring (spring constant = 70 pN/Å) with one end attaching to the C_α atom of PSGL-1 Pro18 or the O1 atom sLe^x GlcNAc and the other end moving at a constant speed of 10 Å/ns. The closed and open angles observed in the crystal structures are marked by a mauve and a blue line, respectively. Other stable conformations are indicated with cyan lines. Data were sampled at 10^{12} Hz. *B*, Force-time courses of SMD simulated unbinding of PSGL-1 from P-selectin lectin and EGF domains. Complete rupture of all preexisting interactions (indicated by an open arrowhead) resulted in a sudden drop in force, but the complex remained bound for ~1 ns of time and sustained ~200 pN of force due to sliding and formation of new interactions. Some long lasting new interactions are shown by color-matched paired arrows, with up arrows indicating their formation and down arrows indicating their dissociation. The residue pairs involved in the new interactions are indicated at the same level as the arrows using matched colors with P-selectin residues on the left and PSGL-1 residues on the right. Final dissociation of the complex is marked by a closed arrowhead. *C*, Number of new interactions observed in 24 SMD simulations vs. PSGL-1 residue involved in new interaction formation. The contributions from various lectin residues are indicated. *Insert*: Mean ± s.e.m. of the number of new interactions observed per simulation is plotted vs. pulling speed. The number of simulations for each of the three pulling speeds is indicated. Reproduced from Ref. [12].

New interactions were observed in all SMD simulations regardless of whether the ligand was PSGL-1 or sLe^x, whether the EGF domain was included or deleted, or whether the EGF domain (when it was included) was unfolded or remained folded.

A total of 109 new interactins were observed in 24 SMD simulations, which are summarized in Fig. 3C. Among the PSGL-1 residues, fucose was most frequently observed to form new interactions. Arg85 was the P-selectin residue seen to form new interactions most frequently. New interaction formation depended on how fast the force was applied (i.e., the loading rate or speed for a fixed spring constant of 70 pN/Å). More new interactions were formed for low speed loading (Fig. 3C insert), suggesting that new interaction formation may be more likely to occur in laboratory experiments where much lower loading speeds are used.

In the simulation with both the lectin and EGF domains, new interaction formation after the initial dissociation was clearly observed to link to the opening of the lectin-EGF interdomain angle. The applied force acted to align the molecular complex along the force direction, which tilted the P-selectin-PSGL-1 binding interface to align more parallel to the force direction after the interdomain angle was opened. The alignment of the binding interface with the force direction provided an opportunity for possible sliding of the ligand on the lectin domain surface. The broad, shallow, and complementary interface likely favors formation of new interactions. In simulations without the EGF domain, the C_α atom of lectin Ala120 was fixed, which also allowed force to align the binding interface, thereby promoting the formation of new interactions.

4 Sliding-Rebinding Mechanism and Pseudoatom Representation

The formation of new interactions after dissociation of preexisting interactions at the atomic level likely prevents unbinding of the P-selectin-PSGL-1 complex at the molecular level, thereby prolonging the lifetimes of the complex. New interaction formation is regulated by the lectin-EGF hinge conformation: It can be promoted by an open-angle conformation but prevented by a close-angle conformation. Free dynamics simulations suggest simultaneous transitions among hinge conformations (Fig. 2B), and SMD simulations suggest forced opening of the hinge angle. In a real system, it is likely that the lectin-EGF hinge conformations are distributed among multiple stable states. The fractions of times during which the molecule would assume each of these conformations are in a dynamic equilibrium, although occasional rapid transitions among the different conformations may occur. Since most crystal structures show a closed angle, it is reasonable to assume that the closed-angle conformation is the most populated conformation in the absence of applied force. In other words, a selectin molecule most likely assumes a closed-angle conformation at zero force. When an external force is applied, the equilibrium contribution is shifted toward open conformations.

Although not directly observed in SMD simulations, it is possible that the formation of new interactions will also enhance the rebinding of the selectin-ligand complex. The longer the molecular pair is kept from complete separation by the newly formed interactions, the more likely is rebinding to occur.

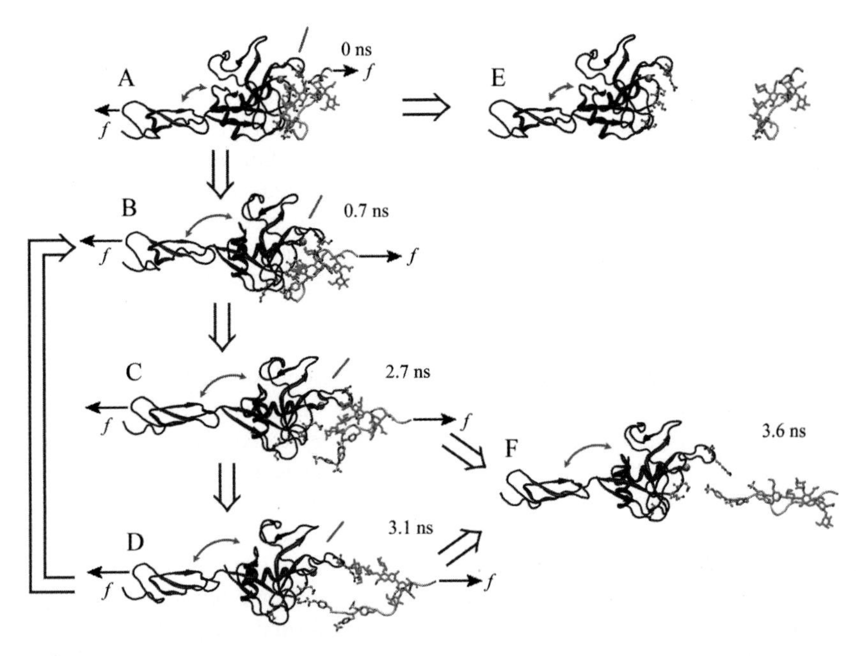

Figure 4. Pathways of the sliding-rebinding mechanism. *A-D* and *F*, Sequential SMD-simulated structures of PSGL-1 (pink for the peptide and green for the glycan) dissociating from P-selectin lectin-EGF domains (blue) at indicated times. The simulation is the same as that shown in Fig. 3B. *E*, Separate structures of P-selectin lectin-EGF domains and PSGL-1 N-terminal fragment indicating complete dissociation. The thick purple arrows indicate the hypothetical sequence of events along dissociation pathways. The interdomain angles are marked by arched double-sided red arrows. The inclinations of the binding interface are marked by inclined red lines. *A*, The initial bound state. *B*, Force-induced opening of the hinge angle. *C*, Rupture of preexisting interactions. *D*, Sliding and formation of new interactions. *E*, Dissociation from fast pathway 1. *F*, Dissociation from slow pathway 2. When a small force *f* (short black arrows) is applied, the complex may detach by dissociation of all noncovalent interactions that preexisted in the bound state. An intermediate force (long black arrows) may open the hinge angle, tilt the binding interface, and promote sliding of PSGL-1 over the lectin binding interface after preexisting atomic-level interactions dissociate. This provides an opportunity for new interactions to form, which would replace those that are disrupted, or for the original interactions to reform, which would return the system back to its previously bound state, before the ligand fully dissociates. Reproduced from Ref. [12].

We developed a sliding-rebinding mechanism to integrate various pieces of information from analyses of the static crystal structures and the dynamic MD-simulated structures, which explains the catch-slip bonds for selectin-ligand interactions [12]. The proposed pathways of the sliding-rebinding mechanism are shown in Fig. 4. At low forces, molecular dissociation is assumed to follow the pathway from Fig. 4A to Fig. 4E. Few new interactions may be formed after initial dissociation of preexisting atomic-level contacts, because most likely the lectin-EGF interdomain angle would remain closed and the binding interface would be relatively perpendicular to the force direction. This would result in short bond lifetimes and fast molecular dissociation. This may be viewed as a directly dissociating or fast dissociation pathway.

Increasing force increases the fraction of time during which the lectin-EGF hinge would assume an open conformation, as shown in Fig. 4B, which would tilt the binding interface towards the force direction. After initial dissociation of preexisting atomic-level interactions (Fig. 4C), the ligand is more likely to slide over the lectin surface and form new interactions (Fig. 4D) or even return to the initial bound state via rebinding (Fig. 4B), which would prolong lifetimes of the molecular bond and give rise to a catch bond behavior. This represents a sliding-rebinding or slow dissociation pathway.

Further increase in force is likely to fully open the interdomain angle and maximize the probability of new interaction formation. No more new interactions can be formed beyond a certain force level. The dissociation of each individual atomic-level interaction, regardless of whether it is preexisting or newly formed, is assumed to be accelerated by increasing force, which, if not countered by new interaction formation and rebinding, gives rise to slip bond behavior at the molecular level. Thus, in the sliding-rebinding model, catch bonds and the catch-slip transition are explained by the competition between the directly dissociating pathway and the sliding-rebinding pathway. The sliding-rebinding pathway is regulated by force through the hinge angle conformation, which controls the probability of new interaction formation.

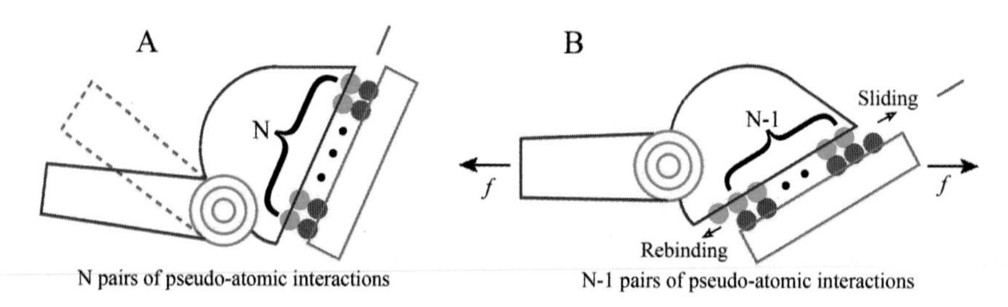

Figure 5. Pseudoatom representation of the sliding-rebinding mechanism for catch bonds. The lectin domain (blue half-ellipse) connects to the EGF domain (blue rectangle for open-angle and mauve dashed rectangle for closed-angle conformations) through an interdomain hinge modeled as a rotational spring (green coiled-coil). The hinge angle conformation is multi-stable and can be regulated by force. The ligand (brown rectangle) interacts with the selectin through pseudoatoms (some shown explicitly as cyan and red circles, others are not shown but implicated by small black dots). The closed-angle system in *A* has N pairs of interacting pseudo-atoms. Force (f, black arrows) induces the opening of the interdomain angle in *B* and induces sliding (shown by dislocation of the cyan and red circles to dock new partners) to a new stage that has N - 1 pairs of interacting pseudoatoms (shown by one red circle from the PSGL-1 moving outside of the interface from the upper-right side, one cyan circle from the lectin moving outside of the interface from the lower-left side, and one less small black dot inside the interface). Further sliding will take the system to the next stage with N - 2 pairs of interacting pseudoatoms. Rebinding will bring the system back to the previous stage with N pairs of interacting pseudoatoms. Reproduced from Ref. [12].

The sliding-rebinding mechanism has been formulated mathematically using reduced representations where noncovalent interactions were modeled as N ($\geq$2) pairs of interacting pseudoatoms (Fig. 5) [12]. The pseudoatomic pairs were assumed to be identical and to interact independently via the same on-rate k_{+1} and off-rate k_{-1}. After all N pairs of pseudoatomic interactions have dissociated (Fig. 5A), the system can either dissociate or slide to a stage with N - 1 pairs of pseudoatomic interactions (Fig. 5B). The probability p_n of sliding and new interaction formation is assumed to depend on force in a manner that resembles the interdomain angle vs. force plot obtained from SMD simulations (Fig. 3A). After all N - 1 pairs of pseudoatomic interactions have dissociated, the system can either dissociate, slide to a stage with N - 2 pseudoatomic interactions, or rebind to the previous stage of N pairs of pseudoatomic interactions with a rebinding rate k_{+2}. Given the number of initial interactions N and the force dependence of parameters k_{-1}, k_{+1}, p_n and k_{+2}, the lifetimes of the molecular bond can be calculated by Monte Carlo simulations.

The validity of the sliding-rebinding mechanism and its pseudoatomic representation has been tested by comparison with experimental data (Fig. 6). Here, k_{-1} is assumed to follow the Bell equation [1], k_{+1} and k_{+2} are assumed to be constants. The possibility of sliding is assumed to be [12]:

$$p_n = \begin{cases} 0 & \text{if } f < 0 \\ \{0.5[1+\sin(\pi f / f_0 - \pi/2)]\}^{1/2} & \text{if } 0 \leq f \leq f_0 \\ 1 & \text{if } f > f_0 \end{cases} \qquad (1)$$

where f_0 is a constant force at which sliding and new contact formation are maximized. As shown in Fig. 6, the model fits the experimental data very well.

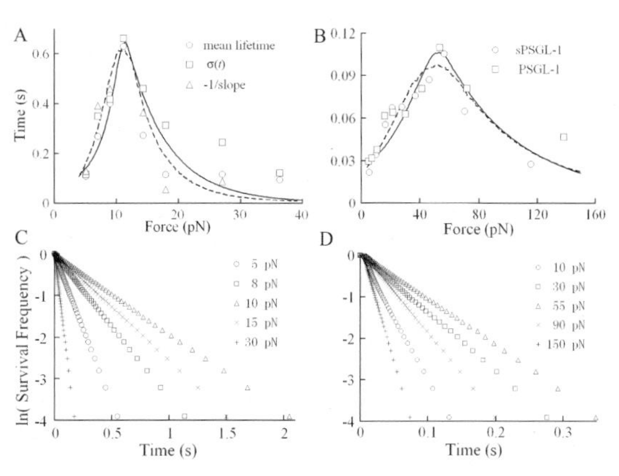

Figure 6. Comparison with experiment. *A* and *B*, Monte Carlo simulated solution of the sliding-rebinding model with 2 (solid curves) or 3 (dashed curves) pairs of pseudoatoms was fit to the lifetime vs. force data of soluble (s) PSGL-1 dissociating from P-selectin (*A*) [6] or of two forms of PSGL-1 dissociating from L-selectin (*B*) [6]. *C* and *D*, Semilog plots of survival frequency vs. time for indicated levels of constant forces. Data were generated using parameters that fitted the bond lifetime vs. force data for P-selectin-sPSGL-1 interaction by the model with two pairs of pseudo-atoms (*C*) or L-selectin-sPSGL-1 interaction by the model with three pairs of pseudo-atoms (*D*). Reproduced from Ref. [12].

5 Testing the Sliding-Rebinding Mechanism by Mutagenesis Studies

Key elements of the sliding-rebinding mechanism for catch bonds include a force-regulated interdomain hinge and a complementary binding interface to enable sliding and new interaction formation. Thus, the validity of the sliding-rebinding mechanism can be tested by mutagenesis studies that change any of these two elements. One of the atomic-level interactions that stabilize the L-selectin lectin-EGF hinge in the closed-angle conformation is a hydrogen bond between Tyr37 and Asn138. Point mutation that replaces Asn by Gly in position 138 (termed L-selectinN138G) would eliminate this hydrogen bond, which should make the interdomain hinge more flexible. According to the sliding-rebinding model, mutant L-selectinN138G should require lower force (smaller f_0 in Eq. 1) to open the lectin-EGF hinge and rebind more readily than wild-type L-selectin. These are predicted to augment catch bonds at the low force regime by lengthening bond lifetimes and lowering the force at which lifetimes reach maximum. Experiments to test these predictions were performed using a biomembrane force probe (BFP) and a flow chamber to measure lifetimes of L-selectin or L-selectinN138G interacting with either PSGL-1 or 6-sulfo-sLe^x, a form of sLe^x with a sulfate ester attached to the C-6 position of GlcNAc [8]. Both experimental method yielded similar results, which agree very well with our predictions (Fig. 7). Like L-selectin, L-selectinN138G interactions with both ligands exhibited transitions between catch and slip bonds. However, the lifetime vs. force curves for L-selectinN138G were shifted leftwards and upwards relative to the L-selectin curves in the catch bond regime. In contrast, there was little difference in the lifetimes of L-selectin and L-selectinN138G interactions in the slip bond regime. Both data of L-selectin and L-selectinN138G interacting with the same ligand (either PSGL-1 or 6-sulfo-sLe^x) can be fitted by

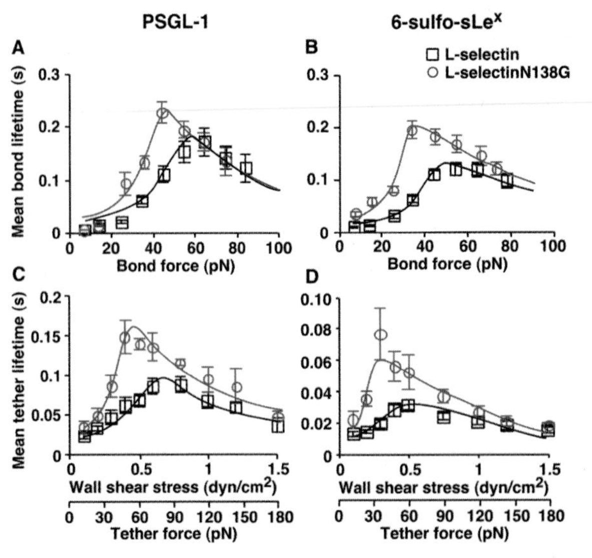

Figure 7. *A* and *C*, Interactions of L-selectin or L-selectinN138G with PSGL-1. *B* and *D*, Interactions of L-selectin or L-selectinN138G with 6-sulfo-sLe^x. Bond lifetimes were measured by BFP (*A* and *B*) or by flow chamber (*C* and *D*) experiments (points), which were fitted by the sliding–rebinding model using Monte Carlo simulations (curves). The data in *A* and *B* represent the mean ± the SEM of ~ 100 lifetime measurements. The data in *C* and *D* represent the mean ± the SD from five experiments. Reproduced from Ref. [8].

the sliding-rebinding model with the same parameters except that f_0 is smaller and k_{+2} is larger for L-selectinN138G than for L-selectin.

6 Discussion and Conclusion

The sliding-rebinding mechanism predicts dissociation along two possible pathways. This is schematically shown Fig. 8 using a hypothetical energy landscape. The force-regulated competition between pathway 1 (directly dissociating or fast dissociation pathway) and pathway 2 (sliding-rebinding or slow dissociation pathway) generates catch bonds at low forces, which transit to slip bonds at high forces. In our sliding-rebinding mechanism, the shifting from pathway 1 to pathway 2 is a consequence of the opening of the force regulated interdomain hinge. Because of this relationship, the mathematical properties of the sliding-rebinding model can be studied in the framework of two-pathway models [11]. However, the sliding-rebinding mechanism suggests a possible structural basis for the two-pathway model for catch bonds. In addition, the sliding-rebinding mechanism provides an explanation for allosteric regulation of molecular interaction where binding is affected by residues far from the binding packet.

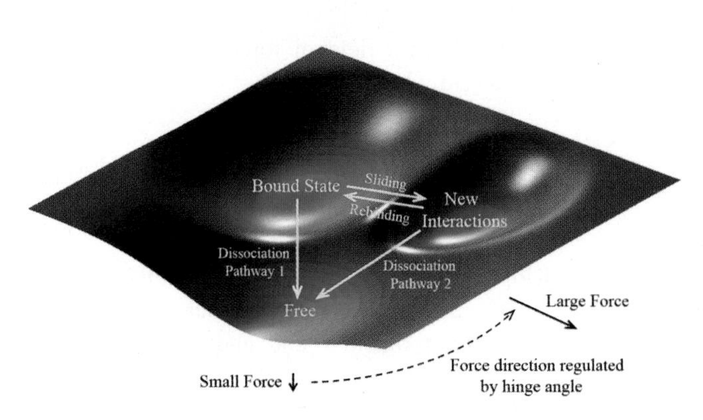

Figure 8. A possible energy landscape for the sliding-rebinding mechanism. Starting from the energy well marked bound state, the molecular complex can dissociate to the free state along pathway 1 or slide into a neighboring energy well resulting from formation of new interactions. From the new interaction energy well the system can dissociate to the free state along pathway 2 or rebind back to the bound-state energy well. As the magnitude of force increases, its direction also changes from that aligned with pathway 1 to that aligned with pathway 2 as the interdomain angle opens by the force, such that sliding into the new interaction energy well increases as force increases from low to intermediate levels. Sliding into the new interaction energy well and rebinding back to the bound-state energy well prolong bond lifetime, thereby giving rise to catch bonds. At high forces all energy barriers are suppressed and dissociation is accelerated in spite of sliding and rebinding, which transitions the catch bonds to slip bonds. Reproduced from Ref. [12].

We should emphasize that the sliding-rebinding mechanism is only one of many possible mechanisms for catch bonds. For example, another possible mechanism may be force induced-fit of the binding interface. Multiple mechanisms may cooperate to produce catch bonds in a single system.

Acknowledgments

We gratefully acknowledge the contributions of our co-workers who produced the original data in Ref. [8]. This work was supported by NIH grant AI44902 (CZ) and HL65631 (RPM).

References

1. Bell, G.I., 1978. Models for the specific adhesion of cells to cells. Science 200, 618-627.
2. Dembo, M., Torney, D.C., Saxman, K., Hammer, D., 1988. The reaction-limited kinetics of membrane-to-surface adhesion and detachment. Proc. R. Soc. Lond. B. Biol. Sci. 234, 55-83.
3. Alon, R., Hammer, D.A., Springer, T.A., 1995. Lifetime of the P-selectin-carbohydrate bond and its response to tensile force in hydrodynamic flow. Nature 374, 539-542.
4. Kramers, H.A., 1940. Brownian motion in a field of force and the diffusion model of chemical reactions. Physica (Utrecht) 7, 284-304.
5. Evans, E. and Ritchie, K., 1997. Dynamic strength of molecular adhesion bonds. Biophys. J. 72, 1541-1555.
6. Marshall, B.T., Long, M., Piper, J.W., Yago, T., McEver, R.P., Zhu, C., 2003. Direct observation of catch bonds involving cell-adhesion molecules. Nature 423, 190-193.
7. Sarangapani, K.K., Yago, T., Klopochi, A.G., Lawrence, M.B., Fieger, C.B., Rosen, S.D., McEver, R.P., Zhu, C., 2004. Low force decelerates L-selectin dissociation from P-selectin glycoprotein ligand-1 and endoglycan. J. Biol. Chem. 279, 2291-2298.
8. Lou, J., Yago, T., Klopochi, A.G., Metha, P., Chen, W., Zarnitsyna, V.I., Bovin, N.V., Zhu, C., McEver, R.P., 2006. Flow-enhanced adhesion regulated by a selectin interdomain hinge. J. Cell. Biol. 174, 1107-1117.
9. Gao, B. and Guilford, W.H., 2006. Mechanics of actomyosin bonds in different nucleotide states are tuned to muscle contraction. Proc. Natl. Acad. Sci. USA 103, 9844-9849.
10. Thomas, W., Forero, M., Yakovenko, O., Nilsson, L., Vicini, P., Sokurenko, E., Vogel, V., 2006. Catch-bond model derived from allostery explains force-activated bacterial adhesion. Biophys. J. 90, 753-764.
11. Zhu, C., Lou, J., McEver, R.P., 2005. Catch bonds: Physical models, structural bases, biological function and rheological relevance. Biorheology 42, 443-462.

12. Lou, J. and Zhu, C., 2007. A structure-based sliding-rebinding mechanism for catch bonds. Biophys. J. in press (doi: 10.1529/biophysj.106.097048).
13. Graves, B.J., Crowther, R.L., Chandran, C., Rumberger, J.M., Li, S., Huang, K.S., Presky, D.H., Familletti, P.C., Wolitzky, B.A., Burns, D.K., 1994. Insight into E-selectin/ligand interaction from the crystal structure and mutagenesis of the lec/EGF domains. Nature 367, 532-538.
14. Somers, W.S., Tang, J., Shaw, G.D., Camphausen, R.T., 2000. Insights into the molecular basis of leukocyte tethering and rolling revealed by structures of P- and E-selectin bound to SLe(X) and PSGL-1. Cell 103, 467-479.
15. Leppänen, A., White, S.P., Helin, J., McEver, R.P., Cummings, R.D., 2000. Binding of glycosulfopeptides to P-selectin requires stereospecific contributions of individual tyrosine sulfate and sugar residues. J. Biol. Chem. 275, 39569-39578.
16. Leppänen, A., Yago, T., Otto, V.I., McEver, R.P., Cummings, R.D., 2003. Model glycosulfopeptides from P-selectin glycoprotein ligand-1 require tyrosine sulfation and a core 2-branched O-glycan to bind to L-selectin. J. Biol. Chem. 278, 26391-26400.
17. Phillips, J.C., Braun, R., Wang, W., Gumbart, J., Tajkhorshid, E., Villa, E., Chipot, C., Skeel, R.D., Kale, L., Schulten, K., 2005. Scalable molecular dynamics with NAMD. J. Comput. Chem. 26, 1781-1802.
18. MacKerell Jr., A.D., Bashford, D., Bellott, M., Dunbrack Jr., R.L., Evanseck, J.D., Field, M.J., Fischer, S., Gao, J., Guo, H., Ha, S., Joseph-McCarthy, D., Kuchnir, L., Kuczera, K., Lau, F.T.K., Mattos, C., Michnick, S., Ngo, T., Nguyen, D.T., Prodhom, B., Reiher, III. W.E., Roux, B., Schlenkrich, M., Smith, J.C., Stote, R., Straub, J., Watanabe, M., Wiorkiewicz-Kuczera, J., Yin, D., Karplus, M., 1998. All-atom empirical potential for molecular modeling and dynamics Studies of proteins. J. Phys. Chem. B. 102, 3586-3616.

ROLE OF EXTERNAL MECHANICAL FORCES IN CELL SIGNAL TRANSDUCTION

*S. R. K. VEDULA AND C. T. LIM

Nano Biomechanics Lab, Division of Bioengineering, National University of Singapore
9 Engineering Drive, Singapore 117576
E-mail: ctlim@nus.edu.sg

*T. S. LIM AND G. RAJAGOPAL

Bioinformatics Institute,
30 Biopolis Street, Matrix, Singapore 138671

W. HUNZIKER

Institute of Molecular and Cell Biology
61 Biopolis Drive, Proteos, Singapore 138673

B. LANE

Centre for Molecular Medicine
61 Biopolis Drive, Proteos, Singapore 138673

M. SOKABE

Department of Physiology, Nagoya University Graduate School of Medicine,
65 Tsurumai, Nagoya, Japan

Mechanotransduction is the process by which cells sense and convert external mechanical stimuli into biochemical signals leading to the regulation of a number of important biological processes such as cell migration, proliferation and differentiation. Though it has been established that many cell types actively respond to mechanical forces, the actual sensing receptors, the signaling pathways activated and the subsequent regulation of biological processes remain largely unclear.

However, recent advances in genomic, proteomic and molecular imaging techniques and technology are providing new insights into the mechanotransduction process. Here we review how cells respond to mechanical forces in their environment, the various "sensors" on the cells for mechanical forces, the different signaling pathways activated, the crosstalk between some of the important pathways and the various "actuators" that determine the biological outcome.

* Both authors contributed equally to the review.

1 Introduction

All types of cells and organisms are continuously subjected to mechanical forces acting on them although the type and magnitude of these forces may differ. For example bone cells are subjected to mostly compressive and tensile forces of large magnitude whereas the endothelial cells lining the blood vessels are predominantly subjected to shear stress and pressure from blood flow which is relatively low in magnitude. However, cells are not "passive absorbers" of these mechanical forces. They have evolved over the years to respond in an organized and favorable way to these forces. For example, external mechanical strain has shown to regulate proliferation, remodeling and gene expression in bone [1-3], cartilage [4-6], epithelial cells [7-9], endothelial cells [10, 11] and fibroblasts [12-15]. This process by which cells sense and respond to external mechanical forces is called mechanotransduction. Though the type and magnitude of response observed may vary from one cell type to another, there are certain underlying unifying principles and processes that are common to the process of mechanotransduction in all types of cells.

The importance of understanding and studying the process of mechanotransduction stems from the fact that a "breakdown" of this process is an important underlying cause of numerous diseases [16]. Examples of such diseases include osteoporosis, congestive heart failure, hypertension and ventilation induced lung injury just to name a few. This chapter aims to give a comprehensive overview of the mechanotransduction process in cells.

1.1 Mechanotransduction process

A simple control loop in engineering systems is depicted in Fig.1 (a). It basically consists of an input stimulus that is sensed by a specialized sensor which in turn conveys it to a controller. Based on the input received, this controller determines the final output or response through an actuator.

If we compare the process of mechanotransduction to such a control loop, the stimulus in this case will be the external mechanical force acting on the cell (Fig.1 (b)). The sensor corresponds to mechanosensor whereas the controller is the complex network of various signaling cascades that are activated in response to the force applied. The actuators are the final downstream molecules of these signaling cascades which regulate the final observable response that can be cell proliferation, orientation, differentiation, migration or apoptosis. Finally, there is some negative feedback response to the controller that would keep a check on the entire process.

The process of mechanotransduction can hence be decomposed into three different processes as shown by the colored boxes in fig. 2 [17]. The red boxes correspond to the "*mechanosensing process,*" the blue boxes constitute the "*mechanotransduction process*" and the green box is the "*mechanoresponse process.*" This chapter aims to provide a general overview of these processes.

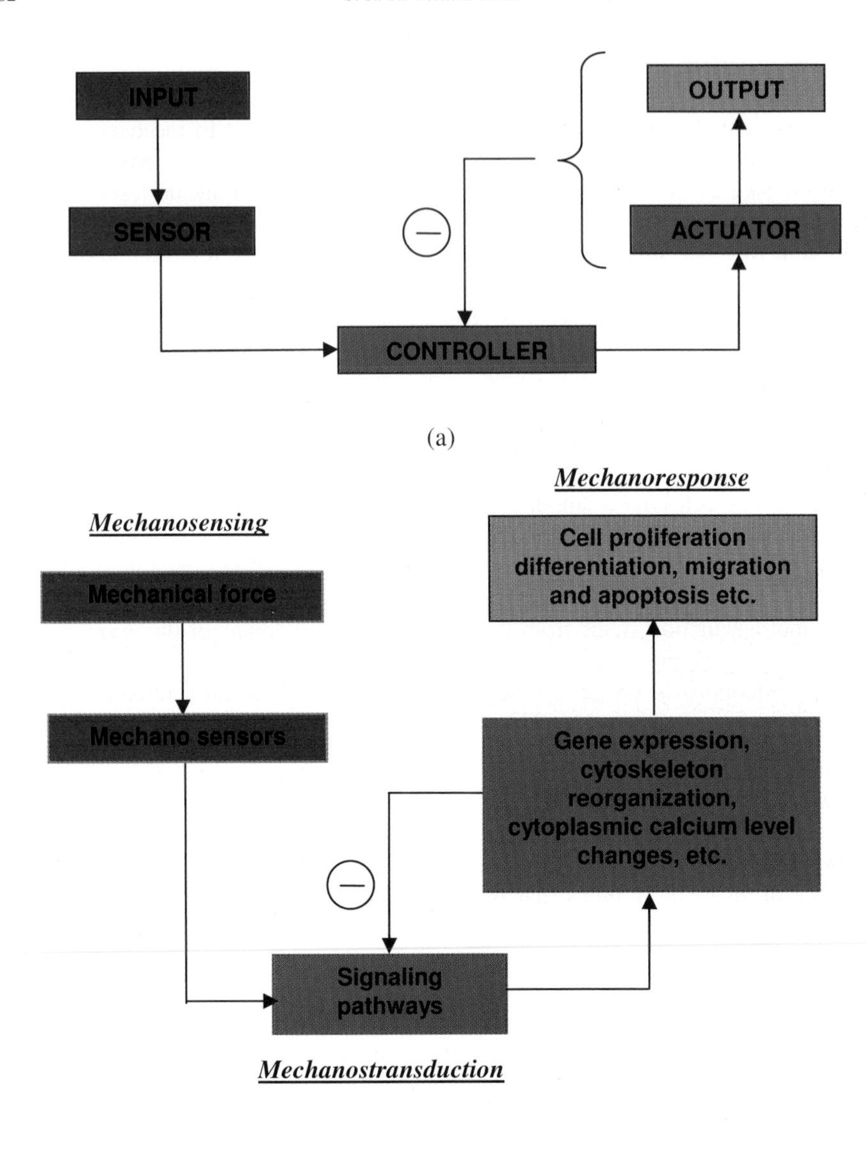

Figure 1. Schematic of (a) simple control loop in engineering systems (b) corresponding analogy to the mechanotransduction process and its three basic components of mechanosensing, mechanotransduction and mechanoresponse [17].

2 Mechanosensing

Mechanosensing is probably the least understood aspect of the whole process of mechanotransduction. Though several mechanosensors have been identified, many questions still remain unanswered. In this section we elaborate on the different mechanosensors and the process by which they sense mechanical forces as we understand now.

2.1 Mechanosensitive or stretch sensitive ion channels

The existence of these channels was first predicted based on the ability of E. coli to resist osmotic shock [18]. Experiments showed that these bacteria had some type of "emergency release valves" that allowed solutes to diffuse out when the bacteria were subjected to hypo-osmotic shock. The search for these "valves" finally led to the discovery of mechanosensitive (MS) ion channels in bacteria [19-21]. There are several excellent reviews describing in detail the structure and function of MS channels [22-25].

(a) Prokaryotic MS channels
There are three types of MS channels: MscM (mini conductance), MscS (small conductance) and MscL (large conductance) [26]. The higher conductance channels need larger pressure for their opening. These probably provide protection at different magnitudes of pressure. Of these, the most well studied and characterized channel is the MscL channel [27, 28]. The solved crystal structure of the MscL of M.*tuberculosis* has helped in understanding the mechanism of opening and closing of these channels. The structure showed that the channel was organized as a homopentamer containing an extracellular pore narrowing onto the intracellular side [29].

(b) Eukaryotic MS channels
Recently several gene products have been identified that probably represent mechanosensitive ion channels in eukaryotic cells. Examples include TREK-1 and TRAAK in mammalian neurons, MEC in C.elegans and TRPC1 in Xenopus and SAKcaC in chick hearts [30-36]. However, they have not yet been characterized or understood in as much details as MS channels in prokaryotes.

2.1.1 Models for the functioning of MS channels

Two types of models have been proposed to describe the function of the MS channels. The first model called the "bilayer model" assumes that "in-plane" tension is sufficient to open these channels. The protein opens up like the iris due to tilting of the transmembrane helices to allow the diffusion of ions when subjected to tension in the membrane plane (Fig. 2a). Support for this model comes from the fact

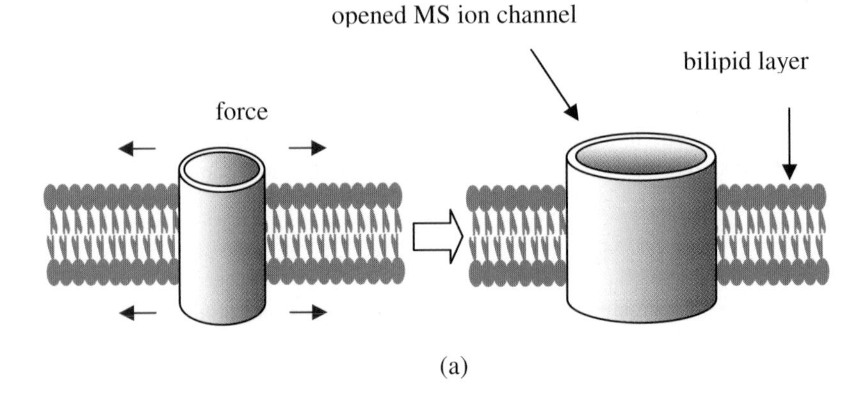

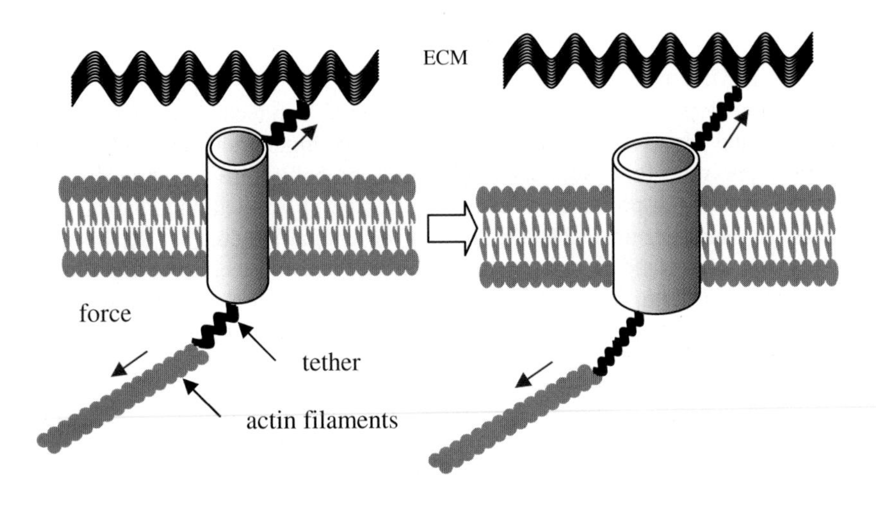

(b)

Figure 2. Schematics of two proposed models for the activation of MS Channels: (a) In the bilayer model, in-plane tension causes the channel protein to open up like an iris (b) In the tethered model, the tethering of the channel to the extracellular matrix proteins and/or cytoskeleton is necessary for the opening of the channel.

that purified MscL protein when reconstituted into lipid bilayers retained its mechanosensitive property [37-41]. The second model requires that the MS channels be tethered to the cytoskeletal elements and the extracellular matrix proteins [42-44]. A relative movement between the anchored ECM proteins and the cytoskeletal elements due to external force could cause the channels to open up (Fig. 2b). Considering the reported values for the stiffness of the membrane [45] and the cytoskeleton [46] cytoskeleton would act as a more efficient force-transmitting and focusing device than membrane owing to its larger elastic modulus [47]. In addition, because of its macroscopic linear structure, cytoskeleton would serve as a force direction antenna as suggested in the mechanotransduction at the hair bundle of the inner ear hair cells [48]. However, further experiments show that plasma membrane vesicles devoid of any cytoskletal components still show MS channel activity though reduced in mechanosensitivity and adaptation to repeated stimuli. This raises the possibility that the cytoskeleton might actually *modulate* the response of MS channels in response to bilayer tension rather than control it.

2.2 Integrins

Integrins are heterodimeric transmembrane cell receptors (made up of monomers of α and β sub-units) that interact with the extracellular matrix proteins on one hand and with the cytoskeleton (through several adaptor molecules like paxillin, vinculin, talin and α-actinin) on the other. This places integrins in an important position for conveying forces between these two elements [49-51]. In fact they have been named integrins because they "integrate" the extracellular matrix proteins and the cytoskeleton.

The initial adhesion of integrins to the ECM proteins called the focal complex leads to their clustering and is later strengthened by recruitment of various kinases (e.g. focal adhesion kinase and Src), adaptor molecules and the cytoskeleton leading to the formation of the mature focal adhesion (FA). The recruitment can regulate the cell cycle via FAK and Fyn/Shc pathways [52-54]. Apart from these main signaling pathways integrins are also necessary for the optimal activation of the signaling pathways associated with other growth factors like insulin, PDGF, EGF and VEGF [55, 56]. Externally applied mechanical forces play an important role in the recruitment of the kinases and adaptor molecules and in the maturation of the FA [57]. This "force dependent stiffening" is a very important characteristic of integrin mediated cell substrate adhesion [58]. Integrins also play an important role in mechanotransduction in endothelial cells in response to shear stress [59-61].

It has been proposed that there are two possible ways in which integrins could act as mechanosensors: the first is the possibility that they open MS channels existing in their vicinity, leading to ion influx and signal initiation [63-65].

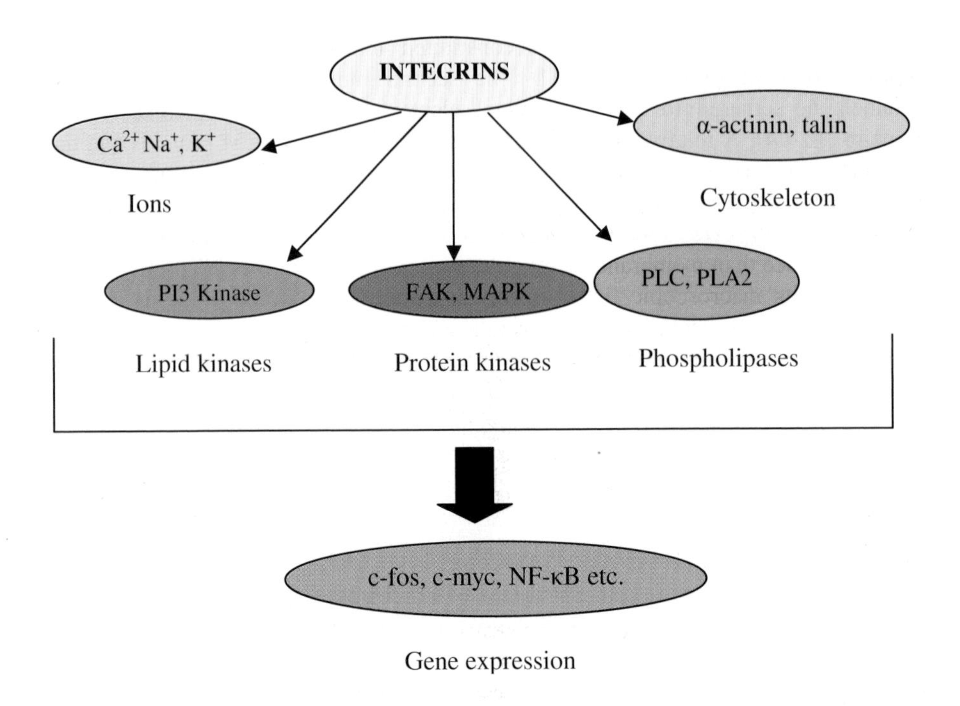

Figure 3. Schematic of signaling pathways activated by integrins [62].

The second possibility is that external force induces a change in the structure and/or relative positions of the large number of molecules associated (kinases and other adaptor molecules) with the cytoplasmic domain of integrins [62, 66-68]. This relative change in the position and/or structure could open up cryptic molecular recognition sites in the proteins that could trigger the cascading of the signals (Fig.3).

2.3 Intercellular adhesion molecules

Apart from shear forces, most other external forces acting on a cell have to be transmitted to it either at the cell substrate interface or the cell-cell interface. It is hence only logical to think that intercellular adhesion molecules could probably play an important role in mechanotransduction at the cell-cell interface. A well developed cell-cell adhesion complex consists of several components that include the adherens junctions (AJ), tight junctions (TJ), gap junctions and desmosomes, which are constituted by cadherins, claudins, connexins, and desmosomal cadherins respectively [69].

Among these, the AJ is perhaps most important for transmitting mechanical signals directly to the actin cytoskeleton. Though there are many similarities between the architecture of FA and AJ, the mechanism of how mechanical forces regulate the formation of AJ and their downstream signaling remains unclear. For example, localization of adaptor protein vinculin at FA was found to correlate with the amount of local tension [70, 71]; it is however unknown whether the localization of vinculin in AJ is similarly induced by tension. It has been shown that application of mechanical force to E-cadherins at well formed intercellular junctions in fibroblasts can cause an increase in cytoplasmic calcium levels [72]. Furthermore, experiments also show that external mechanical strain can lead to phosphorylation of β-catenin in chondrocytes [73]. It has also been shown that the intercellular adhesion molecules VE cadherins and PECAM are important in signal transduction triggered by shear forces in endothelial cells [74, 75] while adhesion of leucocytes to endothelial cells causes phosphorylation of ERK2 [76].

Though current opinion does support the possibility of cadherins acting as mechanosensors, the role of other intercellular adhesion molecules (tight junctions, desmosomes and gap junctions) has not been investigated and provides an exciting area of future research.

2.4 Cytoskeleton

Of the three types of filaments that make up the cytoskeleton (actin, microtubules and intermediate filaments), actin filaments play the most important role in all aspects of mechanotransduction i.e. mechanosensing, mechanotransduction as well as mechanoresponse. The association of actin with cell-matrix and cell-cell adhesion molecules has already been mentioned. This association not only couples cell–ECM receptors such as integrins to the actin–myosin contractile apparatus, but also sequesters signaling molecules that participate in integrin signaling, e.g., focal adhesion kinase (FAK), Shc and Crk [77].

Several mechanisms have been proposed by which structural changes in the cytoskeleton, induced by external forces, could be transduced into biochemical responses [58]. One such mechanism could be the bringing together of an enzyme and its substrate into close proximity. This has been proposed to explain some of the rapidly occurring (over few milliseconds) processes like Ca^{2+} and neurotransmitter release in motor neurons on stressing integrins [78]. Alternatively, the stress in the actin filaments could change the molecular structure of certain proteins associated with actin leading to either opening up of their active sites or change their thermodynamic and kinetic parameters. For example, the actin filament- associated protein (AFAP) can activate c-Src in response to mechanical stretch [79, 80]. Finally, it has also been proposed that applied stress on the cells could be directly conveyed to the nucleus through actin filaments for affecting gene transcription [81-83].

Apart from actin filaments, microtubules are also responsible for the manner in which cells responds to external forces. For example, inhibition of microtubule elongation decreases cell spreading, lamellipodial protrusion, and cell migration [84-89] and stabilization of microtubules with taxol inhibits shear stress-activation of Rac and lamellipodial protrusion in the flow direction [90]. These results suggest that shear stress may induce microtubule elongation in the flow direction, which in turn activates Rac to promote actin polymerization and thus lamellipodial protrusion in the flow direction.

2.5 Other receptors (GPCR and RTK)

Though both G-protein coupled receptors (GPCR) and receptor tyrosine kinases (RTK) are classical receptors activated by the binding of ligands (e.g. growth factors like EGF and PDGF), they have also been shown to be activated by the application of external mechanical force [91-95]. Once activated, these receptors can in turn activate several signaling pathways and regulate gene expression.

2.6 Membrane fluidity

It has been shown in endothelial cells that the fluidity (opposite of viscosity) of the cell membrane changes when fluid shear stress is applied [95, 96]. This change can probably alter the spatial localization or structure of membrane proteins leading to signal initiation. This raises the possibility that the cell membrane itself can act as a mechanosensor for fluid shear stress.

3 Mechanotransduction

The actual signaling cascades that are initiated by the activation of the above described mechanosensors by externally applied mechanical forces are quite complex. Adding to the complexity is the significant amount of crosstalk that occurs between these pathways. In this section, we will focus on some of the important pathways that are initiated by MS channels and integrins.

3.1 Mechanosensitive (MS) ion channels

As previously mentioned, external forces acting on cells can lead to opening up of mechanosensitive ion channels. The influx of ions through these channels can trigger several signaling cascades. Though Na^+ and K^+ ion influx may play an important role in the down stream signaling activated by MS channels in some cases, the most important pathways seem to be mediated by Ca^{2+} and as discovered recently, by MS release of chemical transmitters, notably ATP.

3.1.1 Elevation of intracellular calcium levels

Calcium plays an extremely important role in regulating several cellular processes such as cell growth, proliferation, differentiation and migration. It is therefore not surprising that Ca^{2+} entry through MS ion channels is one of the most important processes in mechanotransduction. The increase in intracellular Ca^{2+} in response to mechanical strain has been shown in many cell types including smooth muscle cells [97], cardiomyocytes [98], fibroblasts [99], osteoblasts [100], hair cells [101], neurons [102], epithelial cells [103] and vascular endothelial cells (VECs) [104]. Ca^{2+} entry through MS Ca^{2+} channels can be amplified by triggering Ca^{2+}-induced Ca^{2+} release [105], activation of phospholipase C activity and inositol 1,4,5-trisphosphate (IP3)-sensitive Ca^{2+} release [103, 106, 107]. Even in the complete absence of external Ca^{2+}, mechanical stimulation can elevate Ca^{2+} by promoting Ca^{2+} release from internal stores, mostly the endoplasmic reticulum. The versatility of Ca^{2+} signaling is greatly enhanced by the numerous cross talks that link it to other signaling pathways. A few important pathways that are mediated by an increase in cytoplasmic Ca^{2+} levels are shown in Fig.4. There are several reviews focused on a more detailed description of Ca^{2+} mediated signaling pathways [108-110].

3.1.2 Mechanosensitive (MS) ATP Release

Though the major focus of research in mechanotransduction has been on MS ion channels, it has recently become clear that MS release of other transmitters, most notably ATP, plays an important role especially in eukaryotic cells. In *Xenopus* oocyte, it has been shown that MS ATP release is more sensitive to mechanical stimuli than MS ion channels [111-113]. Furthermore, MS ATP release has also been implicated in vertebrate touch and stretch sensation [112, 114], processes which are commonly assumed to be mediated by MS ion channels. The MS release of ATP may subsequently induce an electrical or biochemical signal by activating purinergic family of receptors [115], which either constitute an integral part of the ion channels [116-118] or are indirectly coupled to membrane ion channels via soluble second messenger pathways [119, 120].

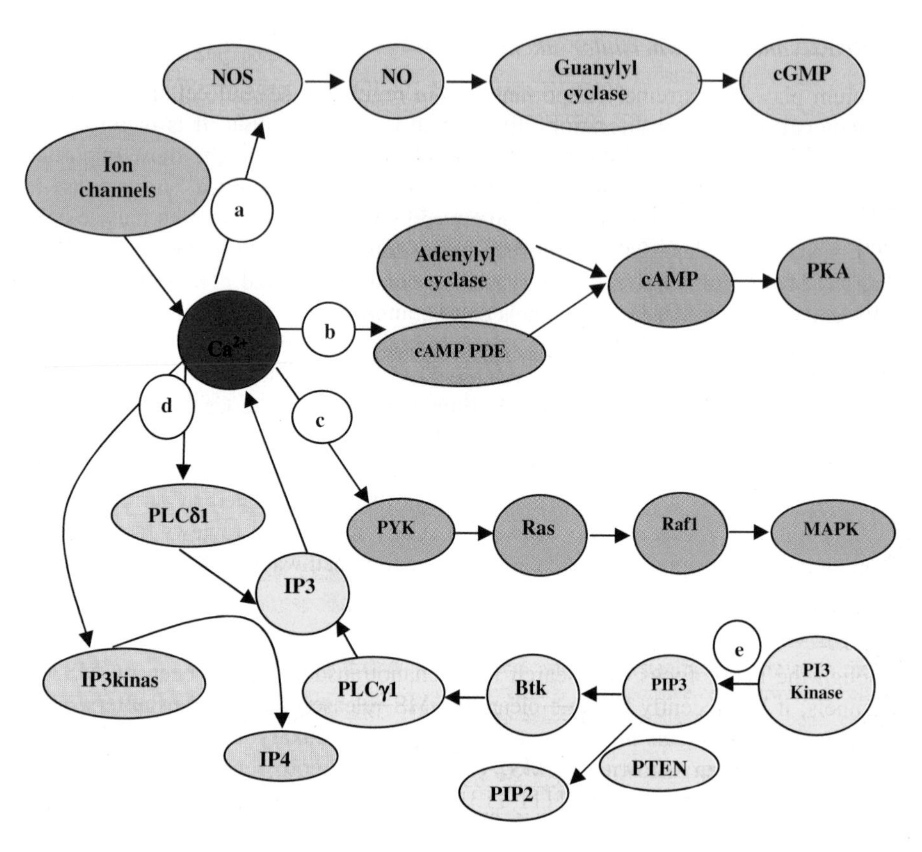

Figure 4. Schematic of signaling pathways initiated by intracellular influx of Ca^{2+} [108] (a) Ca^{2+}induced Nitric Oxide (NO) synthesis. NO synthase is induced by Ca^{2+} to generate NO, which activates guanylyl cyclase to produce cyclic GMP (cGMP) (b) Ca^{2+}–cyclic AMP interactions. Ca^{2+} affects the cellular cAMP levels because it can influence the activity of both adenylyl cyclase (that increases cAMP) and cAMP phosphodiesterases (cAMP PDE that hydrolyses cAMP) (c) Ca^{2+}–mitogen-activated protein kinase interaction. Mitogen-activated protein kinase (MAPK) signaling pathway can be triggered by Ca^{2+} by activating proline-rich tyrosine kinase 2 (PYK2) and Ras (d) Ca^{2+} feedback interactions. There are many feedback interactions within the Ca^{2+} signaling pathway whereby Ca^{2+} can modulate its own activity. For example, Ca^{2+}can reduce Ins(1,4,5)P3 by activating Ins(1,4,5)P3 kinase to produce Ins(1,3,4,5)P4. Conversely, it can activate phospholipase Cδ1 (PLCδ1) to increase the level of Ins(1,4,5)P3. (e) Ca^{2+}–phosphatidylinositol-3-OH kinase interaction Phosphatidylinositol-3,4,5-trisphosphate (PtdIns(3,4,5)P3) acts as a second messenger to affect phosphatidylinositol-3-OH-kinase (PI(3)K) signaling pathway by activating non-receptor tyrosine kinase Btk. Btk phosphorylates and activates phospholipase Cγ1 (PLCγ1) [121], which further influences Ca^{2+} through Ins(1,4,5)P3. The tumor suppressor PTEN, a 3-phosphatase that lowers the level of PtdIns(3,4,5)P3, reduces both the level of Ins(1,4,5)P3 and the influx of external Ca^{2+} [122].

3.2 Integrins, Focal Complex (FC) & Focal Adhesion (FA)

As described before integrins can transmit mechanical stress bidirectionally between the membrane and ECM. The transition from initial interaction between integrins and ECM to FC requires the recruitment of adaptor proteins like paxillin, talin and vinculin to connect the FC to actin cytoskeleton [57, 123]. FC are precursors of FA [124], which develop in response to Rho activation. Increasing force at adhesion sites triggers the clustering of integrins, the recruitment of adhesion structural proteins, and the elongation of the adhesion site in the direction of force [70, 125, 126]. Forces applied to ECM–integrin–cytoskeleton connections, which can be generated by internal actin or external ECM motion, induce maturation of adhesion sites to FA, which are coupled to bundles of actin called stress fibers [70, 126]. Conversely, loss of force triggers the disassembly of stress fibers and adhesion sites [70].

Integrins are the starting points for several signaling pathways important of which are the FAK and the Fyn/Shc pathways (Fig. 5). Integrin mediated increase in intracellular Ca^{2+}, probably through MS channels, has already been mentioned before.

3.2.1 FAK pathway

FAK or focal adhesion kinase is a tyrosine kinase that is well known to respond to integrin clustering and FA formation by auto-phosphorylation and activation [127]. Auto-phoshphorylation of FAK on Tyr^{397} provides a binding site for SH2 domain of Src, Fyn and PI3-kinase. The phosphorylated Src and Fyn can in turn phosphorylate several adaptor molecules like paxillin and docking proteins like $p130^{CAS}$. Recent studies show that applying mechanical stretch can increase FAK phosphorylation, sustain the activation of the mitogen-activated protein kinase (MAPK) ERK, and cause cells to proliferate [128, 129]. Furthermore, cyclic strain has been shown to activate c-Src which can phosphorylate Tyr^{925} of FAK. This provides a binding site for the SH2 domain of Grb. The association of SOS can then lead to activation of Ras and ERK pathway [130, 131]. Furthermore, the phosphorylation of $p130^{CAS}$ recruits Crk that can activate another MAPK (Jun N-terminal kinase; JNK). Activated JNK can then translocate into the nucleus and phosphorylate c-jun leading to transcription of genes essential for cell proliferation.

3.2.2 Fyn/Shc pathway

Ras-ERK pathway is one of the most important pathways activated by receptor tyrosine kinases usually after the binding of growth factors. However, mechanical stretch acting through integrins can also activate this pathway in several ways. One way of MAPK activation as described before is through phosphorylation of FAK that can then bind to the adaptor Grb. The second way in which MAPK can be activated through integrin activation is by the Fyn/Shc pathway. Fyn is a tyrosine kinase that can be activated only by certain types of integrins and probably needs caveolin-1 as a membrane adapter. SH3 domain on activated Fyn can bind to proline rich residues in adapter protein Shc and phosphorylate it. This provides a binding site for Grb-SOS adapter complex to activate Ras and then MAPK [132]. However, some responses to mechanical strain are cell-type and ECM-type specific. For example, static biaxial stretch on cardiac fibroblasts stimulate ERK2 in cells plated on fibronectin; activates JNK1 in cells placed on substrate coated with vitronectin or laminin; but does not activate any MAPK pathway in cells plated on collagen [133].

3.2.3 Rho family GTPases

Rho family of GTPases (Cdc42, Rac and Rho) constitutes a group of molecules that regulate the actin based cytoskeletal structure of the cell and hence play an important role in deciding cell morphology and migration. They are responsible for the formation of filopodia, lamellipodia and stress fibres respectively [134]. Regulation of these GTPases partly through integrin mediated signaling hence plays an important role in deciding as to how external forces influence cell shape and migration. The effect of external mechanical force on these small GTPases is complex and probably depends on the type of cell. For example, it has been shown that external cyclical strain can inhibit [135] or activate Rac [136]. Also studies show that cells not only tend to migrate faster on rigid substrates but they also tend to migrate towards areas of the substrate with higher rigidity [137]. Rho and Rac probably play an important role in this process (durotaxis). Apart from their direct role in influencing cell migration and morphology, they also influence other cellular activities by their ability to cross talk with other signaling pathways e.g. by regulating activation of ERK [138].

3.2.4 Tyrosine phosphatases

Tyrosine phosphatases play an important role in controlling both FAK and Fyn mediated signaling pathways. Some phosphatases like RPTP-α and SHP-2 amplify the FAK and Fyn/Shc pathway by dephosphorylating the inhibitory sites in Src-family kinases [139] or by preventing the phosphorylation and inhibition of α-actinin by FAK [140]. Other phosphatases like PTP-1B and PTEN can inhibit these pathways by dephosphorylating and inhibiting $p130^{CAS}$ or FAK and Shc respectively [141, 142]. Accordingly, drugs that can inhibit tyrosine phosphatases have a profound but varied effect on adhesion-site formation [143]. For example, the non-specific tyrosine phosphatase inhibitor phenyl arsine oxide (PAO) induces the formation of stable adhesion sites on soft substrates [137] but inhibits adhesion reinforcement triggered by restrained force applied to fibronectin-coated beads [144].

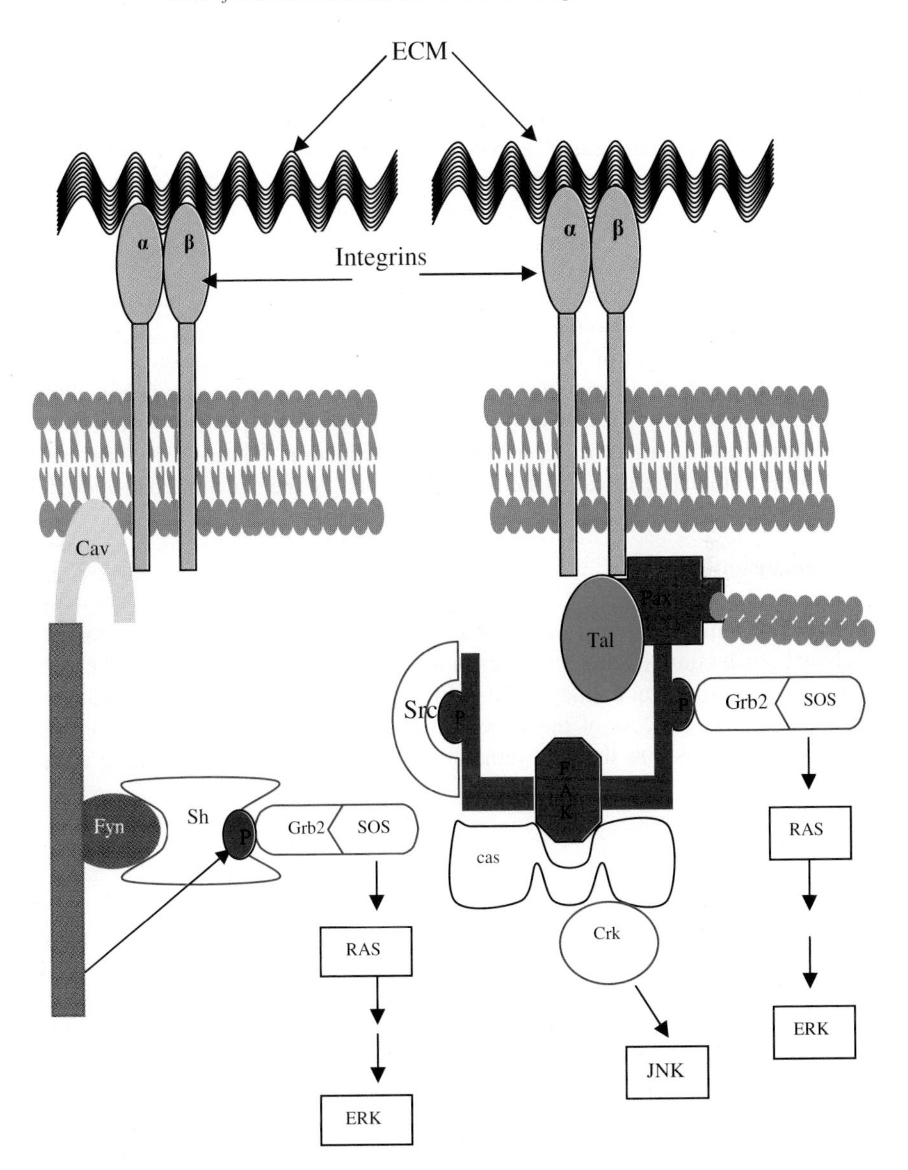

Figure 5. Schematic of FAK and Fyn/Shc pathways at focal adhesion complex leading to ERK and JNK activation. Initial binding occurs between integrins and the extracellular matrix proteins followed by recruitment of several adaptor molecules and enzymes on the cytoplasmic end of the integrins [54].

In conclusion, integrins and the associated adapter proteins located in FA play a very important role in mechanotransduction. However, at this juncture it is extremely difficult to pinpoint on the exact contribution of each molecule to the whole process of mechanotransduction.

4 Mechanoresponse

Mechanoresponse or the visible outcome of application of external mechanical force to cells depends not only on the type of cell but also on the duration, magnitude and frequency of application of forces. Though many signaling pathways are activated by the application of external forces on cells, however, there are only a limited number of observable responses like gene expression, protein secretion, cell morphology, proliferation, differentiation and migration. These responses are the result of the cumulative effect of the final "effectors" of the signaling pathways.

5 Conclusion

This review aims to give a simplified and general overview of the process of mechanotransduction. Recent advances in nano-technological techniques like optical traps and imaging modalities like GFP tagging have greatly helped us in understanding the process of mechanotransduction in much more detail. However, there are many questions that still remain unanswered and provide exciting avenues for future research.

References

1. Tang, L., Lin, Z.Li, Y.M., 2006. Effects of different magnitudes of mechanical strain on Osteoblasts in vitro, Biochem. Biophys. Res. Commun. 344, 122-128.
2. Kusumi, A., Sakaki, H., Kusumi, T., Oda, M., Narita, K., Nakagawa, H., Kubota, K., Satoh, H.Kimura, H., 2005. Regulation of synthesis of osteoprotegerin and soluble receptor activator of nuclear factor-kappaB ligand in normal human osteoblasts via the p38 mitogen-activated protein kinase pathway by the application of cyclic tensile strain, J. Bone Miner. Metab. 23, 373-381.
3. Kaspar, D., Seidl, W., Neidlinger-Wilke, C.Claes, L., 2000. In vitro effects of dynamic strain on the proliferative and metabolic activity of human osteoblasts, J. Musculoskelet. Neuronal Interact. 1, 161-164.
4. Ng, T.C., Chiu, K.W., Rabie, A.B.Hagg, U., 2006. Repeated mechanical loading enhances the expression of Indian hedgehog in condylar cartilage, Front. Biosci. 11, 943-948.

5. Lahiji, K., Polotsky, A., Hungerford, D.S.Frondoza, C.G., 2004. Cyclic strain stimulates proliferative capacity, alpha2 and alpha5 integrin, gene marker expression by human articular chondrocytes propagated on flexible silicone membranes, In Vitro Cell Dev Biol Anim. 40, 138-142.
6. Li, K.W., Williamson, A.K., Wang, A.S.Sah, R.L., 2001. Growth responses of cartilage to static and dynamic compression, Clin. Orthop. Relat. Res. S34-48.
7. Takei, T., Rivas-Gotz, C., Delling, C.A., Koo, J.T., Mills, I., McCarthy, T.L., Centrella, M.Sumpio, B.E., 1997. Effect of strain on human keratinocytes in vitro, J Cell Physiol. 173, 64-72.
8. Han, O., Li, G.D., Sumpio, B.E.Basson, M.D., 1998. Strain induces Caco-2 intestinal epithelial proliferation and differentiation via PKC and tyrosine kinase signals, Am. J. Physiol. 275, G534-541.
9. Li, W., Duzgun, A., Sumpio, B.E.Basson, M.D., 2001. Integrin and FAK-mediated MAPK activation is required for cyclic strain mitogenic effects in Caco-2 cells, Am. J. Physiol Gastrointest Liver Physiol. 280, G75-87.
10. Nishimura, K., Li, W., Hoshino, Y., Kadohama, T., Asada, H., Ohgi, S.Sumpio, B.E., 2006. Role of AKT in cyclic strain-induced endothelial cell proliferation and survival, Am. J. Physiol. Cell. Physiol. 290, C812-821.
11. Murata, K., Mills, I.Sumpio, B.E., 1996. Protein phosphatase 2A in stretch-induced endothelial cell proliferation, J. Cell. Biochem. 63, 311-319.
12. Zeichen, J., van Griensven, M.Bosch, U., 2000. The proliferative response of isolated human tendon fibroblasts to cyclic biaxial mechanical strain, Am. J. Sports. Med. 28, 888-892.
13. Berry, C.C., Cacou, C., Lee, D.A., Bader, D.L.Shelton, J.C., 2003. Dermal fibroblasts respond to mechanical conditioning in a strain profile dependent manner, Biorheology. 40, 337-345.
14. Danciu, T.E., Gagari, E., Adam, R.M., Damoulis, P.D.Freeman, M.R., 2004. Mechanical strain delivers anti-apoptotic and proliferative signals to gingival fibroblasts, J. Dent. Res. 83, 596-601.
15. Grunheid, T.Zentner, A., 2005. Extracellular matrix synthesis, proliferation and death in mechanically stimulated human gingival fibroblasts in vitro, Clin. Oral. Investig. 9, 124-130.
16. Ingber, D.E., 2003. Mechanobiology and diseases of mechanotransduction, Ann. Med. 35, 564-577.
17. Vogel, V.Sheetz, M., 2006. Local force and geometry sensing regulate cell functions, Nat. Rev. Mol. Cell Biol. 7, 265-275.
18. Britten, R.J.Mc, C.F., 1962. The amino acid pool in Escherichia coli, Bacteriol. Rev. 26, 292-335.
19. Berrier, C., Coulombe, A., Szabo, I., Zoratti, M.Ghazi, A., 1992. Gadolinium ion inhibits loss of metabolites induced by osmotic shock and large stretch-activated channels in bacteria, Eur. J. Biochem. 206, 559-565.

20. Martinac, B., Buechner, M., Delcour, A.H., Adler, J.Kung, C., 1987. Pressure-sensitive ion channel in Escherichia coli, Proc. Natl. Acad. Sci. USA 84, 2297-2301.
21. Yang, X.C. Sachs, F., 1989. Block of stretch-activated ion channels in Xenopus oocytes by gadolinium and calcium ions, Science. 243, 1068-1071.
22. Kung, C., 2005. A possible unifying principle for mechanosensation, Nature. 436, 647-654.
23. Martinac, B., 2004. Mechanosensitive ion channels: molecules of mechanotransduction, J. Cell Sci. 117, 2449-2460.
24. Perozo, E., 2006. Gating prokaryotic mechanosensitive channels, Nat. Rev. Mol. Cell Biol. 7, 109-119.
25. Wood, J.M., 1999. Osmosensing by bacteria: signals and membrane-based sensors, Microbiol. Mol. Biol. Rev. 63, 230-262.
26. Berrier, C., Besnard, M., Ajouz, B., Coulombe, A.Ghazi, A., 1996. Multiple mechanosensitive ion channels from Escherichia coli, activated at different thresholds of applied pressure, J. Membr. Biol. 151, 175-187.
27. Sukharev, S.I., Blount, P., Martinac, B., Blattner, F.R. Kung, C., 1994. A large-conductance mechanosensitive channel in E. coli encoded by mscL alone, Nature. 368, 265-268.
28. Sukharev, S.I., Blount, P., Martinac, B.Kung, C., 1997. Mechanosensitive channels of Escherichia coli: the MscL gene, protein, and activities, Annu. Rev. Physiol. 59, 633-657.
29. Chang, G., Spencer, R.H., Lee, A.T., Barclay, M.T.Rees, D.C., 1998. Structure of the MscL homolog from Mycobacterium tuberculosis: a gated mechanosensitive ion channel, Science. 282, 2220-2226.
30. Patel, A.J., Honore, E., Maingret, F., Lesage, F., Fink, M., Duprat, F. Lazdunski, M., 1998. A mammalian two pore domain mechano-gated S-like K+ channel, Embo J. 17, 4283-4290.
31. Patel, A.J., Lazdunski, M.Honore, E., 2001. Lipid and mechano-gated 2P domain K(+) channels, Curr. Opin. Cell Biol. 13, 422-428.
32. Patel, A.J.Honore, E., 2001. Properties and modulation of mammalian 2P domain K+ channels, Trends Neurosci. 24, 339-346.
33. Maingret, F., Patel, A.J., Lesage, F., Lazdunski, M.Honore, E., 2000. Lysophospholipids open the two-pore domain mechano-gated K(+) channels TREK-1 and TRAAK, J. Biol. Chem. 275, 10128-10133.
34. Hamill, O.P.McBride, D.W., Jr., 1996. A supramolecular complex underlying touch sensitivity, Trends Neurosci. 19, 258-261.
35. Chelur, D.S., Ernstrom, G.G., Goodman, M.B., Yao, C.A., Chen, L., R, O.H. Chalfie, M., 2002. The mechanosensory protein MEC-6 is a subunit of the C. elegans touch-cell degenering channel, Nature. 420, 669-673.

36. Tang, Q.Y., Qi, Z., Naruse, K.Sokabe, M., 2003. Characterization of a functionally expressed stretch-activated BKca channel cloned from chick ventricular myocytes, J. Membr. Biol. 196, 185-200.
37. Martinac, B., Adler, J.Kung, C., 1990. Mechanosensitive ion channels of E. coli activated by amphipaths, Nature. 348, 261-263.
38. Sukharev, S., Betanzos, M., Chiang, C.S.Guy, H.R., 2001. The gating mechanism of the large mechanosensitive channel MscL, Nature. 409, 720-724.
39. Sukharev, S., Durell, S.R.Guy, H.R., 2001. Structural models of the MscL gating mechanism, Biophys. J. 81, 917-936.
40. Perozo, E., Cortes, D.M., Sompornpisut, P., Kloda, A. Martinac, B., 2002. Open channel structure of MscL and the gating mechanism of mechanosensitive channels, Nature. 418, 942-948.
41. Perozo, E., Kloda, A., Cortes, D.M.Martinac, B., 2002. Physical principles underlying the transduction of bilayer deformation forces during mechanosensitive channel gating, Nat. Struct. Biol. 9, 696-703.
42. Hamill, O.P.McBride, D.W., Jr., 1997. Induced membrane hypo/hyper-mechanosensitivity: a limitation of patch-clamp recording, Annu. Rev. Physiol. 59, 621-631.
43. Hamill, O.P.Martinac, B., 2001. Molecular basis of mechanotransduction in living cells, Physiol. Rev. 81, 685-740.
44. Zhang, Y., Gao, F., Popov, V.L., Wen, J.W.Hamill, O.P., 2000. Mechanically gated channel activity in cytoskeleton-deficient plasma membrane blebs and vesicles from Xenopus oocytes, J. Physiol. 523 Pt 1, 117-130.
45. Evans, E.A., Waugh, R.Melnik, L., 1976. Elastic area compressibility modulus of red cell membrane, Biophys. J. 16, 585-595.
46. Kojima, H., Ishijima, A.Yanagida, T., 1994. Direct measurement of stiffness of single actin filaments with and without tropomyosin by in vitro nanomanipulation, Proc. Natl. Acad. Sci. USA 91, 12962-12966.
47. Sokabe, M., Sachs, F.Jing, Z.Q., 1991. Quantitative video microscopy of patch clamped membranes stress, strain, capacitance, and stretch channel activation, Biophys. J. 59, 722-728.
48. Pickles, J.O.Corey, D.P., 1992. Mechanoelectrical transduction by hair cells, Trends Neurosci. 15, 254-259.
49. Wang, N., Butler, J.P.Ingber, D.E., 1993. Mechanotransduction across the cell surface and through the cytoskeleton, Science. 260, 1124-1127.
50. Salter, D.M., Robb, J.E.Wright, M.O., 1997. Electrophysiological responses of human bone cells to mechanical stimulation: evidence for specific integrin function in mechanotransduction, J. Bone Miner. Res. 12, 1133-1141.
51. Katsumi, A., Orr, A.W., Tzima, E.Schwartz, M.A., 2004. Integrins in mechanotransduction, J. Biol. Chem. 279, 12001-12004.
52. Schwartz, M.A., Schaller, M.D.Ginsberg, M.H., 1995. Integrins: emerging paradigms of signal transduction, Annu. Rev. Cell. Dev. Biol. 11, 549-599.

53. Shyy, J.Y. Chien, S., 1997. Role of integrins in cellular responses to mechanical stress and adhesion, Curr. Opin. Cell Biol. 9, 707-713.
54. Giancotti, F.G. Ruoslahti, E., 1999. Integrin signaling, Science. 285, 1028-1032.
55. Vuori, K.Ruoslahti, E., 1994. Association of insulin receptor substrate-1 with integrins, Science. 266, 1576-1578.
56. Miyamoto, S., Teramoto, H., Gutkind, J.S.Yamada, K.M., 1996. Integrins can collaborate with growth factors for phosphorylation of receptor tyrosine kinases and MAP kinase activation: roles of integrin aggregation and occupancy of receptors, J. Cell Biol. 135, 1633-1642.
57. Galbraith, C.G., Yamada, K.M.Sheetz, M.P., 2002. The relationship between force and focal complex development, J. Cell Biol. 159, 695-705.
58. Ingber, D.E., 1997. Tensegrity: the architectural basis of cellular mechanotransduction, Annu. Rev. Physiol. 59, 575-599.
59. Girard, P.R.Nerem, R.M., 1995. Shear stress modulates endothelial cell morphology and F-actin organization through the regulation of focal adhesion-associated proteins, J. Cell Physiol. 163, 179-193.
60. Takahashi, M.Berk, B.C., 1996. Mitogen-activated protein kinase (ERK1/2) activation by shear stress and adhesion in endothelial cells. Essential role for a herbimycin-sensitive kinase, J. Clin. Invest. 98, 2623-2631.
61. Shyy, J.Y.Chien, S., 2002. Role of integrins in endothelial mechanosensing of shear stress, Circ. Res. 91, 769-775.
62. Shakibaei, M.Mobasheri, A., 2003. Beta1-integrins co-localize with Na, K-ATPase, epithelial sodium channels (ENaC) and voltage activated calcium channels (VACC) in mechanoreceptor complexes of mouse limb-bud chondrocytes, Histol. Histopathol. 18, 343-351.
63. Browe, D.M.Baumgarten, C.M., 2003. Stretch of beta 1 integrin activates an outwardly rectifying chloride current via FAK and Src in rabbit ventricular myocytes, J. Gen. Physiol. 122, 689-702.
64. Wu, Z., Wong, K., Glogauer, M., Ellen, R.P.McCulloch, C.A., 1999. Regulation of stretch-activated intracellular calcium transients by actin filaments, Biochem. Biophys. Res. Commun. 261, 419-425.
65. Meredith, J.E., Jr., Winitz, S., Lewis, J.M., Hess, S., Ren, X.D., Renshaw, M.W.Schwartz, M.A., 1996. The regulation of growth and intracellular signaling by integrins, Endocr. Rev. 17, 207-220.
66. Katsumi, A., Naoe, T., Matsushita, T., Kaibuchi, K.Schwartz, M.A., 2005. Integrin activation and matrix binding mediate cellular responses to mechanical stretch, J. Biol. Chem. 280, 16546-16549.
67. Chen, H.C., Appeddu, P.A., Parsons, J.T., Hildebrand, J.D., Schaller, M.D.Guan, J.L., 1995. Interaction of focal adhesion kinase with cytoskeletal protein talin, J. Biol. Chem. 270, 16995-16999.

68. Lewis, J.M. Schwartz, M.A., 1995. Mapping in vivo associations of cytoplasmic proteins with integrin beta 1 cytoplasmic domain mutants, Mol. Biol. Cell. 6, 151-160.
69. Tsukita, S., Furuse, M. Itoh, M., 2001. Multifunctional strands in tight junctions, Nat. Rev. Mol. Cell Biol. 2, 285-293.
70. Balaban, N.Q., Schwarz, U.S., Riveline, D., Goichberg, P., Tzur, G., Sabanay, I., Mahalu, D., Safran, S., Bershadsky, A., Addadi, L.Geiger, B., 2001. Force and focal adhesion assembly: a close relationship studied using elastic micropatterned substrates, Nat. Cell Biol. 3, 466-472.
71. Tan, J.L., Tien, J., Pirone, D.M., Gray, D.S., Bhadriraju, K.Chen, C.S., 2003. Cells lying on a bed of microneedles: an approach to isolate mechanical force, Proc. Natl. Acad. Sci. USA 100, 1484-1489.
72. Ko, K.S., Arora, P.D.McCulloch, C.A., 2001. Cadherins mediate intercellular mechanical signaling in fibroblasts by activation of stretch-sensitive calcium-permeable channels, J. Biol. Chem. 276, 35967-35977.
73. Lee, H.S., Millward-Sadler, S.J., Wright, M.O., Nuki, G.Salter, D.M., 2000. Integrin and mechanosensitive ion channel-dependent tyrosine phosphorylation of focal adhesion proteins and beta-catenin in human articular chondrocytes after mechanical stimulation, J. Bone Miner. Res. 15, 1501-1509.
74. Tzima, E., Irani-Tehrani, M., Kiosses, W.B., Dejana, E., Schultz, D.A., Engelhardt, B., Cao, G., DeLisser, H. Schwartz, M.A., 2005. A mechanosensory complex that mediates the endothelial cell response to fluid shear stress, Nature. 437, 426-431.
75. Osawa, M., Masuda, M., Kusano, K.Fujiwara, K., 2002. Evidence for a role of platelet endothelial cell adhesion molecule-1 in endothelial cell mechanosignal transduction: is it a mechanoresponsive molecule?, J. Cell Biol. 158, 773-785.
76. Cuvelier, S.L., Paul, S., Shariat, N., Colarusso, P. Patel, K.D., 2005. Eosinophil adhesion under flow conditions activates mechanosensitive signaling pathways in human endothelial cells, J. Exp. Med. 202, 865-876.
77. Critchley, D.R., 2000. Focal adhesions - the cytoskeletal connection, Curr. Opin. Cell Biol. 12, 133-139.
78. Chen, B.M.Grinnell, A.D., 1995. Integrins and modulation of transmitter release from motor nerve terminals by stretch, Science. 269, 1578-1580.
79. Lodyga, M., Bai, X.H., Mourgeon, E., Han, B., Keshavjee, S.Liu, M., 2002. Molecular cloning of actin filament-associated protein: a putative adaptor in stretch-induced Src activation, Am. J. Physiol. Lung Cell Mol. Physiol. 283, L265-274.
80. Han, B., Bai, X.H., Lodyga, M., Xu, J., Yang, B.B., Keshavjee, S., Post, M. Liu, M., 2004. Conversion of mechanical force into biochemical signaling, J. Biol. Chem. 279, 54793-54801.

81. Maniotis, A.J., Chen, C.S.Ingber, D.E., 1997. Demonstration of mechanical connections between integrins, cytoskeletal filaments, and nucleoplasm that stabilize nuclear structure, 10.1073/pnas.94.3.849, PNAS. 94, 849-854.
82. Pienta, K.J.Coffey, D.S., 1992. Nuclear-cytoskeletal interactions: evidence for physical connections between the nucleus and cell periphery and their alteration by transformation, J. Cell. Biochem. 49, 357-365.
83. Hu, S., Chen, J., Butler, J.P.Wang, N., 2005. Prestress mediates force propagation into the nucleus, Biochem. Biophys. Res. Commun. 329, 423-428.
84. Gotlieb, A.I., Subrahmanyan, L.Kalnins, V.I., 1983. Microtubule-organizing centers and cell migration: effect of inhibition of migration and microtubule disruption in endothelial cells, J. Cell Biol. 96, 1266-1272.
85. Domnina, L.V., Rovensky, J.A., Vasiliev, J.M.Gelfand, I.M., 1985. Effect of microtubule-destroying drugs on the spreading and shape of cultured epithelial cells, J. Cell Sci. 74, 267-282.
86. Bershadsky, A.D., Vaisberg, E.A.Vasiliev, J.M., 1991. Pseudopodial activity at the active edge of migrating fibroblast is decreased after drug-induced microtubule depolymerization, Cell Motil. Cytoskeleton. 19, 152-158.
87. Mikhailov, A.Gundersen, G.G., 1998. Relationship between microtubule dynamics and lamellipodium formation revealed by direct imaging of microtubules in cells treated with nocodazole or taxol, Cell Motil. Cytoskeleton. 41, 325-340.
88. Waterman-Storer, C.M., Worthylake, R.A., Liu, B.P., Burridge, K. Salmon, E.D., 1999. Microtubule growth activates Rac1 to promote lamellipodial protrusion in fibroblasts, Nat. Cell Biol. 1, 45-50.
89. Wittmann, T., Bokoch, G.M.Waterman-Storer, C.M., 2003. Regulation of leading edge microtubule and actin dynamics downstream of Rac1, J. Cell Biol. 161, 845-851.
90. Hu, Y.L., Li, S., Miao, H., Tsou, T.C., del Pozo, M.A. Chien, S., 2002. Roles of microtubule dynamics and small GTPase Rac in endothelial cell migration and lamellipodium formation under flow, J. Vasc. Res. 39, 465-476.
91. Correa-Meyer, E., Pesce, L., Guerrero, C.Sznajder, J.I., 2002. Cyclic stretch activates ERK1/2 via G proteins and EGFR in alveolar epithelial cells, Am. J. Physiol. Lung Cell Mol Physiol. 282, L883-891.
92. Chen, K.D., Li, Y.S., Kim, M., Li, S., Yuan, S., Chien, S.Shyy, J.Y., 1999. Mechanotransduction in response to shear stress. Roles of receptor tyrosine kinases, integrins, and Shc, J. Biol. Chem. 274, 18393-18400.
93. Gudi, S., Huvar, I., White, C.R., McKnight, N.L., Dusserre, N., Boss, G.R.Frangos, J.A., 2003. Rapid activation of Ras by fluid flow is mediated by Galpha(q) and Gbetagamma subunits of heterotrimeric G proteins in human endothelial cells, Arterioscler. Thromb. Vasc. Biol. 23, 994-1000.
94. Labrador, V., Chen, K.D., Li, Y.S., Muller, S., Stoltz, J.F.Chien, S., 2003. Interactions of mechanotransduction pathways, Biorheology. 40, 47-52.

95. Gudi, S., Nolan, J.P.Frangos, J.A., 1998. Modulation of GTPase activity of G proteins by fluid shear stress and phospholipid composition, Proc. Natl. Acad. Sci. USA 95, 2515-2519.
96. Haidekker, M.A., L'Heureux, N.Frangos, J.A., 2000. Fluid shear stress increases membrane fluidity in endothelial cells: a study with DCVJ fluorescence, Am. J. Physiol. Heart Circ. Physiol. 278, H1401-1406.
97. Kirber, M.T., Guerrero-Hernandez, A., Bowman, D.S., Fogarty, K.E., Tuft, R.A., Singer, J.J.Fay, F.S., 2000. Multiple pathways responsible for the stretch-induced increase in Ca2+ concentration in toad stomach smooth muscle cells, J. Physiol. 524 Pt 1, 3-17.
98. Sigurdson, W., Ruknudin, A. Sachs, F., 1992. Calcium imaging of mechanically induced fluxes in tissue-cultured chick heart: role of stretch-activated ion channels, Am. J. Physiol. 262, H1110-1115.
99. Bibby, K.J.McCulloch, C.A., 1994. Regulation of cell volume and [Ca2+]i in attached human fibroblasts responding to anisosmotic buffers, Am. J. Physiol. 266, C1639-1649.
100. Pommerenke, H., Schreiber, E., Durr, F., Nebe, B., Hahnel, C., Moller, W.Rychly, J., 1996. Stimulation of integrin receptors using a magnetic drag force device induces an intracellular free calcium response, Eur. J. Cell Biol. 70, 157-164.
101. Hudspeth, A.J.Gillespie, P.G., 1994. Pulling springs to tune transduction: adaptation by hair cells, Neuron. 12, 1-9.
102. Chen, Y., Simasko, S.M., Niggel, J., Sigurdson, W.J.Sachs, F., 1996. Ca2+ uptake in GH3 cells during hypotonic swelling: the sensory role of stretch-activated ion channels, Am. J. Physiol. 270, C1790-1798.
103. Boitano, S., Sanderson, M.J.Dirksen, E.R., 1994. A role for Ca(2+)-conducting ion channels in mechanically-induced signal transduction of airway epithelial cells, J. Cell Sci. 107 (Pt 11), 3037-3044.
104. Demer, L.L., Wortham, C.M., Dirksen, E.R. Sanderson, M.J., 1993. Mechanical stimulation induces intercellular calcium signaling in bovine aortic endothelial cells, Am. J. Physiol. 264, H2094-2102.
105. Naruse, K.Sokabe, M., 1993. Involvement of stretch-activated ion channels in Ca2+ mobilization to mechanical stretch in endothelial cells, Am. J. Physiol. 264, C1037-1044.
106. Brophy, C.M., Mills, I., Rosales, O., Isales, C.Sumpio, B.E., 1993. Phospholipase C: a putative mechanotransducer for endothelial cell response to acute hemodynamic changes, Biochem. Biophys. Res. Commun. 190, 576-581.
107. Matsumoto, H., Baron, C.B.Coburn, R.F., 1995. Smooth muscle stretch-activated phospholipase C activity, Am. J. Physiol. 268, C458-465.
108. Berridge, M.J., Lipp, P.Bootman, M.D., 2000. The versatility and universality of calcium signalling, Nat. Rev. Mol. Cell Biol. 1, 11-21.

109. Webb, S.E.Miller, A.L., 2003. Calcium signalling during embryonic development, Nat. Rev. Mol. Cell Biol. 4, 539-551.
110. Berridge, M.J., Bootman, M.D.Roderick, H.L., 2003. Calcium signalling: dynamics, homeostasis and remodelling, Nat. Rev. Mol. Cell Biol. 4, 517-529.
111. Maroto, R.Hamill, O.P., 2001. Brefeldin A block of integrin-dependent mechanosensitive ATP release from Xenopus oocytes reveals a novel mechanism of mechanotransduction, J. Biol. Chem. 276, 23867-23872.
112. Nakamura, F.Strittmatter, S.M., 1996. P2Y1 purinergic receptors in sensory neurons: contribution to touch-induced impulse generation, Proc. Natl. Acad. Sci. USA 93, 10465-10470.
113. Zhang, Y.Hamill, O.P., 2000. On the discrepancy between whole-cell and membrane patch mechanosensitivity in Xenopus oocytes, J. Physiol. 523 Pt 1, 101-115.
114. Cook, S.P., Vulchanova, L., Hargreaves, K.M., Elde, R.McCleskey, E.W., 1997. Distinct ATP receptors on pain-sensing and stretch-sensing neurons, Nature. 387, 505-508.
115. North, R.A.Barnard, E.A., 1997. Nucleotide receptors, Curr. Opin. Neurobiol. 7, 346-357.
116. Brehm, P., Kullberg, R.Moody-Corbett, F., 1984. Properties of non-junctional acetylcholine receptor channels on innervated muscle of Xenopus laevis, J Physiol. 350, 631-648.
117. Chen, C.C., Akopian, A.N., Sivilotti, L., Colquhoun, D., Burnstock, G.Wood, J.N., 1995. A P2X purinoceptor expressed by a subset of sensory neurons, Nature. 377, 428-431.
118. Lewis, C., Neidhart, S., Holy, C., North, R.A., Buell, G.Surprenant, A., 1995. Coexpression of P2X2 and P2X3 receptor subunits can account for ATP-gated currents in sensory neurons, Nature. 377, 432-435.
119. Lustig, K.D., Shiau, A.K., Brake, A.J.Julius, D., 1993. Expression cloning of an ATP receptor from mouse neuroblastoma cells, Proc. Natl. Acad. Sci. USA 90, 5113-5117.
120. Mosbacher, J., Maier, R., Fakler, B., Glatz, A., Crespo, J.Bilbe, G., 1998. P2Y receptor subtypes differentially couple to inwardly-rectifying potassium channels, FEBS Lett. 436, 104-110.
121. Akagi, K., Nagao, T.Urushidani, T., 1999. Correlation between Ca(2+) oscillation and cell proliferation via CCK(B)/gastrin receptor, Biochim. Biophys. Acta. 1452, 243-253.
122. Morimoto, A.M., Tomlinson, M.G., Nakatani, K., Bolen, J.B., Roth, R.A.Herbst, R., 2000. The MMAC1 tumor suppressor phosphatase inhibits phospholipase C and integrin-linked kinase activity, Oncogene. 19, 200-209.
123. DePasquale, J.A.Izzard, C.S., 1991. Accumulation of talin in nodes at the edge of the lamellipodium and separate incorporation into adhesion plaques at focal contacts in fibroblasts, J. Cell Biol. 113, 1351-1359.

124. Rottner, K., Hall, A.Small, J.V., 1999. Interplay between Rac and Rho in the control of substrate contact dynamics, Curr. Biol. 9, 640-648.
125. Ballestrem, C., Hinz, B., Imhof, B.A.Wehrle-Haller, B., 2001. Marching at the front and dragging behind: differential alphaVbeta3-integrin turnover regulates focal adhesion behavior, J. Cell Biol. 155, 1319-1332.
126. Riveline, D., Zamir, E., Balaban, N.Q., Schwarz, U.S., Ishizaki, T., Narumiya, S., Kam, Z., Geiger, B.Bershadsky, A.D., 2001. Focal contacts as mechanosensors: externally applied local mechanical force induces growth of focal contacts by an mDia1-dependent and ROCK-independent mechanism, J. Cell Biol. 153, 1175-1186.
127. Schlaepfer, D.D., Hauck, C.R.Sieg, D.J., 1999. Signaling through focal adhesion kinase, Prog. Biophys. Mol. Biol. 71, 435-478.
128. Numaguchi, K., Eguchi, S., Yamakawa, T., Motley, E.D.Inagami, T., 1999. Mechanotransduction of rat aortic vascular smooth muscle cells requires RhoA and intact actin filaments, Circ. Res. 85, 5-11.
129. Domingos, P.P., Fonseca, P.M., Nadruz, W., Jr.Franchini, K.G., 2002. Load-induced focal adhesion kinase activation in the myocardium: role of stretch and contractile activity, Am. J. Physiol Heart Circ. Physiol. 282, H556-564.
130. Yano, Y., Geibel, J.Sumpio, B.E., 1996. Tyrosine phosphorylation of pp125FAK and paxillin in aortic endothelial cells induced by mechanical strain, Am. J. Physiol. 271, C635-649.
131. Sai, X., Naruse, K.Sokabe, M., 1999. Activation of pp60(src) is critical for stretch-induced orienting response in fibroblasts, J. Cell Sci. 112 (Pt 9), 1365-1373.
132. Wary, K.K., Mainiero, F., Isakoff, S.J., Marcantonio, E.E.Giancotti, F.G., 1996. The adaptor protein Shc couples a class of integrins to the control of cell cycle progression, Cell. 87, 733-743.
133. MacKenna, D.A., Dolfi, F., Vuori, K.Ruoslahti, E., 1998. Extracellular signal-regulated kinase and c-Jun NH2-terminal kinase activation by mechanical stretch is integrin-dependent and matrix-specific in rat cardiac fibroblasts, J. Clin. Invest. 101, 301-310.
134. Nobes, C.D.Hall, A., 1995. Rho, rac, and cdc42 GTPases regulate the assembly of multimolecular focal complexes associated with actin stress fibers, lamellipodia, and filopodia, Cell. 81, 53-62.
135. Katsumi, A., Milanini, J., Kiosses, W.B., del Pozo, M.A., Kaunas, R., Chien, S., Hahn, K.M.Schwartz, M.A., 2002. Effects of cell tension on the small GTPase Rac, J. Cell Biol. 158, 153-164.
136. Kumar, A., Murphy, R., Robinson, P., Wei, L.Boriek, A.M., 2004. Cyclic mechanical strain inhibits skeletal myogenesis through activation of focal adhesion kinase, Rac-1 GTPase, and NF-kappaB transcription factor, Faseb J. 18, 1524-1535.

137. Pelham, R.J., Jr.Wang, Y., 1997. Cell locomotion and focal adhesions are regulated by substrate flexibility, Proc. Natl. Acad. Sci. USA 94, 13661-13665.
138. Laboureau, J., Dubertret, L., Lebreton-De Coster, C.Coulomb, B., 2004. ERK activation by mechanical strain is regulated by the small G proteins rac-1 and rhoA, Exp. Dermatol. 13, 70-77.
139. Oh, E.S., Gu, H., Saxton, T.M., Timms, J.F., Hausdorff, S., Frevert, E.U., Kahn, B.B., Pawson, T., Neel, B.G.Thomas, S.M., 1999. Regulation of early events in integrin signaling by protein tyrosine phosphatase SHP-2, Mol. Cell Biol. 19, 3205-3215.
140. von Wichert, G., Haimovich, B., Feng, G.S.Sheetz, M.P., 2003. Force-dependent integrin-cytoskeleton linkage formation requires downregulation of focal complex dynamics by Shp2, Embo J. 22, 5023-5035.
141. Liu, F., Sells, M.A.Chernoff, J., 1998. Protein tyrosine phosphatase 1B negatively regulates integrin signaling, Curr. Biol. 8, 173-176.
142. Gu, J., Tamura, M.Yamada, K.M., 1998. Tumor suppressor PTEN inhibits integrin- and growth factor-mediated mitogen-activated protein (MAP) kinase signaling pathways, J. Cell Biol. 143, 1375-1383.
143. Schoenwaelder, S.M.Burridge, K., 1999. Bidirectional signaling between the cytoskeleton and integrins, Curr. Opin. Cell Biol. 11, 274-286.
144. Sawada, Y.Sheetz, M.P., 2002. Force transduction by Triton cytoskeletons, J. Cell Biol. 156, 609-615.

III. TISSUE ENGINEERING

EVALUATION OF MATERIAL PROPERTY OF TISSUE-ENGINEERED CARTILAGE BY MAGNETIC RESONANCE IMAGING AND SPECTROSCOPY

S. MIYATA

Graduate School of Life Science and System Engineering, Kyushu Institute of Technology,
2-4 Wakamatsu, Kitakyushu 808-0196, Japan
E-mail: miyata@life.kyutech.ac.jp

K. HOMMA

National Institute of Advanced Industrial Science and Technology, Japan

T. NUMANO

Tokyo Metropollitan University, Japan

K. FURUKAWA AND T. USHIDA

Graduate School of Medicine, University of Tokyo

T. TATEISHI

National Institute for Material Science, Japan

Recently, many kinds of methodology to regenerate articular cartilage have been developed. Applying tissue-engineered cartilage in a clinical setting requires non- or low-invasive evaluation to detect the maturity of cartilage tissue. However there are few studies about noninvasive or nondestructive evaluation of tissue-engineered cartilage. In this study, we established the methodologies of noninvasive evaluation of tissue maturity using magnetic resonance imaging (MRI) and nondestructive evaluation of molecular structure using magnetic resonance spectroscopy (MRS), and performed these evaluations on tissue-engineered cartilage by chondrocyte-seeded agarose culture model. As the results, the sulphated glycosaminoglycan content of the tissue-engineered cartilage showed significant correlation with fixed charge density measured by MRI. The 1H-NMR spectrum of the tissue-engineered cartilage showed peaks dominated similarly to chondroitin sulfate solution. In conclusion, these methodologies using MRI and MRS could be useful for the evaluation of tissue-engineered cartilage.

1 Introduction

Articular cartilage is avascular tissue covering articulating surfaces of bones, and it functions to bear loads and reduce friction in diarthrodial joints. It is a porous gel of large proteoglycan aggregates containing high fixed charge density (FCD) embedded in a water-swollen network of collagen fibrils [1, 2].

Although articular cartilage may function well over a lifetime, traumatic injury or the degenerative changes associated with osteoarthritis (OA) can significantly erode the articular layer, leading to joint pain and instability [3]. Because of its avascular nature, articular cartilage has a very limited capacity to regenerate and repair. Moreover, the natural response of articular cartilage to injury is said to be variable and, at best, unsatisfactory. Therefore, numerous studies have reported tissue-engineering approaches to restore degenerated cartilage and repair defects; these approaches involve culturing autologous chondrocytes *in vitro* to create three-dimensional tissue that is subsequently implanted [4–7]. Clinical application of these approaches requires a noninvasive method of assessing the maturity of the actual engineered tissue to be used therapeutically, and the method should be applicable to various aspects of cartilage regenerative medicine, including characterization of regenerated tissue during *in vitro* culture and *in vivo* evaluation after transplantation.

Magnetic resonance imaging (MRI) and nuclear magnetic resonance (NMR) spectroscopy of articular cartilage is well accepted [8–11] and has been applied in recent years. In addition, the highly promising area of MRI to assess cartilage biochemistry is under investigation [12,13]. Indeed, specific material properties of cartilage explants have already been correlated with specific MR imaging parameters. For instance, increased water content in degenerated cartilage has been correlated with increased self-diffusion of water [14], and loss of proteoglycan has been correlated with MRI-determined FCD using gadolinium diethylene-triaminepentaacetic acid (Gd-DTPA^{2-}) as a contrast agent [15].

Although the noninvasive assessment of tissue maturity and the nondestructive evaluation of molecular structure are important, we believe no previous study has fully evaluated the relationships between the biochemical properties and MRI and NMR measurements of regenerated cartilage consisting of articular chondrocytes. Previous study has indicated that MR images of autologous chondrocyte transplants may show clinically significant variations [16]; neither biochemical properties nor the FCD of regenerated articular cartilage has been evaluated.

In this study, we established nondestructive method to evaluate the molecular structure of tissue-engineered cartilage using NMR spectroscopy. Moreover, we tested the hypothesis that gadolinium-enhanced MR images of tissue-engineered cartilage correlate with biochemical properties and that these novel approaches can be used to assess cartilaginous matrix material properties during tissue maturation.

2 Materials and Methods

2.1 Isolation of chondrocytes and preparation of chondrocyte/agarose constructs

Articular chondrocytes were obtained from the glenohumeral joint of 4- to 6-week-old calves from a local abattoir. Articular cartilage was excised from the humeral head, diced into pieces of ~1mm^3, and then shaken gently in Dulbecco's modified Eagle's medium (DMEM)/Ham's F12 (F12) supplemented with 5% fetal bovine serum (FBS), 0.2% collagenase type II, and antibiotics-antimycotics for 8 hours at 37°C. The digested solution was passed through a 70-μm nylon mesh filter to get rid of debris. Cells were then isolated from the digest by centrifugation for 5 minutes and rinsed twice with phosphate buffered saline (PBS). Finally, after centrifugation for 5 minutes, the cells were resuspended with feed medium (DMEM/F12 supplemented with FBS, 50 μg/mL L-ascorbic acid, and antibiotics-antimycotics), and the total number of cells was counted by hemocytometer.

Chondrocyte-suspended agarose gels were prepared as described previously [17]. Briefly, isolated chondrocytes in the feed medium were mixed with an equal volume of PBS containing low melting temperature agarose (Agarose type VII, Sigma, MO) at 37°C to prepare 1.5×10^7 cells/mL in 2% (wt/wt) agarose gel and cast into a custom-made mold. After gelling at 4°C for 25 minutes, approximately 50 disks of 8-mm diameter, 1.5-mm thickness were cored from the large gel plate with a biopsy punch. The chondrocyte/agarose disks were fed 2.5 mL feed medium/disk per every other day and maintained in an atmosphere of 5% CO_2 at 37°C.

2.2 *1H-NMR spectroscopy*

NMR data were acquired on a Varian NMR400 (Varian, CA, USA) at 400MHz. Spectra were obtained for a $\pi/2$ pulse and a relaxation delay time of 2 s. All measurements were carried out at room temperature 23°C using the deuterated water (D_2O) as a deuterium lock for field stabilization. For each sample spectrum 64 transients were accumulated.

NMR measurements were performed on the chondrocyte-seeded agarose gel (cultured for 2 days and 42 days), 2% (wt/wt) agarose gel, and chondroitin sulfate solution. The chondrocyte-seeded agarose specimens and the 2% agarose gel were soaked two times and equilibrated in deuterated water containing 0.9% (wt/wt) NaCl to minimize a signal of water protons (H_2O and HOD). To establish physiological condition, the deuterated water contained NaCl at 0.9% (wt/wt) which is the same concentration of saline. To set the gel specimen at a center of probe coil and give a sufficient deuterium signal for NMR system, the specimen were put into the NMR glass tube and adjusted the position by 2% (wt/wt) agarose gel made by deuterated water (Fig. 1) [18]. The sodium salts of shark chondroitin sulfate (CS) were dissolved in deuterated water to give a final concentration of 8% (wt/wt) and were allowed to equilibrate for 1 day at 4°C.

2.3 *FCD measurements by Gd-DTPA^{2-} enhanced MRI*

MRI measurements were performed with a 2.0-Tesla Bruker Biospec 20/30 system as described previously [19]. In all MRI measurements, the specimens were put into glass tubes filled with PBS (Fig. 2). We used gadolinium diethylene-triaminepentaacetic acid (Gd-DTPA^{2-}) as a negatively charged contrast agent.

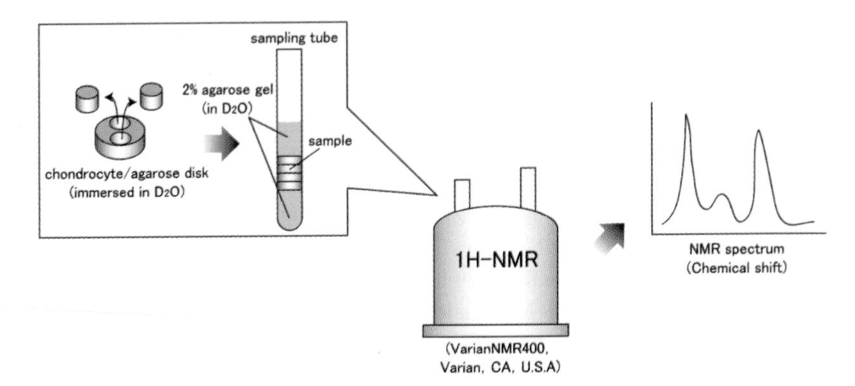

Figure 1. Schematic diagram of NMR spectroscopy of chondrocyte/agarose constructs and 2% agarose gel. Solid specimens were adjusted the position by 2% agarose gel spacer made by deuterated water. [18]

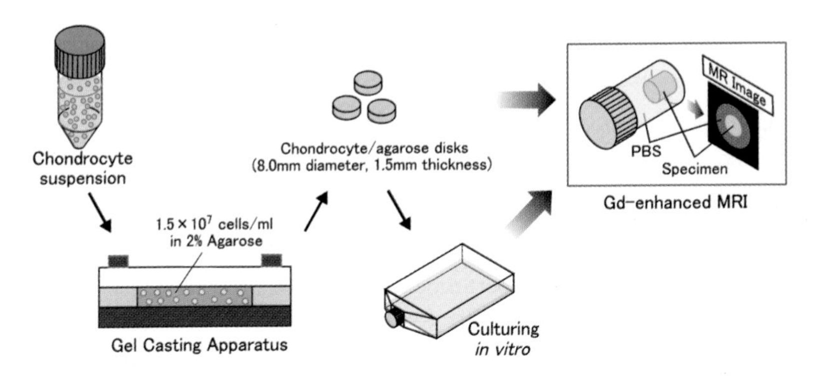

Figure 2. Schematic diagram of agarose gel culture and gadolinium-enhanced MRI. Chondrocyte-seeded agarose gels were made with chondrocyte suspension and agarose solution. In all MRI measurements, the cultured specimens were put into glass tubes filled with phosphate buffered saline (PBS). [19]

The longitudinal relaxation time map, T1-map, was obtained with a short-echo time (TE: 15 ms), spin-echo sequence with different repetition time values (TR: 100 ms to 15 s, 16 steps). Subsequently, the specimens were balanced in PBS containing 1 mM Gd-$DTPA^{2-}$ for 12 hours; the longitudinal relaxation time map in the contrast agent, $T1_{Gd}$-map, was obtained again with a short-echo time (TE: 15 ms), spin-echo sequence with different repetition time values (TR: 30 ms to 5 s, 13 steps). Finally, using the relaxivity (R) value of Gd-$DTPA^{2-}$ in saline (5.24 in our MRI system), the concentration of the contrast agent was estimated using the formula $[\text{Gd-DTPA}^{2-}] = 1/R(1/T1_{Gd} - 1/T1)$. The negative fixed charge density (FCD) was calculated as follows

$$\mathrm{FCD} = \frac{[\mathrm{Na}^+]_b\sqrt{[\mathrm{Gd-DTPA}^{2-}]_t}}{\sqrt{[\mathrm{Gd-DTPA}^{2-}]_b}} - \frac{[\mathrm{Na}^+]_b\sqrt{[\mathrm{Gd-DTPA}^{2-}]_b}}{\sqrt{[\mathrm{Gd-DTPA}^{2-}]_t}}$$

where subscript b stands for bath solution and subscript t stands for cartilaginous tissue [15]. All MRI measurements were performed at room temperature 23°C.

2.4 Histological analysis

Each week, the chondrocyte/agarose disks were removed from culture and processed for histological analysis. Before fixation, they were rinsed twice in PBS. The specimens were fixed in 2% (v/v) glutaraldehyde solution buffered with 0.05 M sodium cacodylate for 1 hour and washed twice with PBS and soaked in 20% sucrose for 24 hours before snap freezing. The fixed samples were made into cryosectins (5 to 7 μm in thickness) and stained with Safranin O. The stained sections were photographed with a phase-contrast microscope.

2.5 Determination of total sulfated glycosaminoglycan contents

Each week, the chondrocyte/agarose disks were also removed from the culture to determine the total sulfated glycosaminoglycan (sGAG) contents and MRI measurement simultaneously. The specimens were weighed wet, lyophilized, weighed dry, and digested with 125 μg/mL papain in 0.1 M sodium acetate containing 0.05 M EDTA and 0.01 M cysteine-HCl at 60°C for 16 hours. Sulfated GAG content of the digests was determined using the 1,9-dimethylmethylene blue (DMMB) dye-binding assay [20], as modified for microtiter plates, and using absorbance at 620 nm. Shark chondroitin sulfate (Chondroitin 6-sulfate, Sigma-Aldrich, MO) was used as the standard (12.5 to 100 μg/ml). Pearson's correlation analysis was applied to assess the linear relationships between MRI and the biochemical parameters.

3 Results

In the 1H-NMR measurements, peaks were found at 2.0 ppm (Fig.3a) and 3.0-4.5 ppm (Fig.3b) in the aqueous chondroitin sulfate spectrum. In the spectrum of the chondrocyte-seeded agarose cultured for 42 days, one narrow peak was also found at 2.0 ppm (Fig. 4b) and narrow peaks were found at 3.0-4.5 ppm (Fig. 4a) and imposed by a broad peak. The chondrocyte-seeded agarose cultured for 2 days and 2% agarose gel showed almost same spectra and one broad peak was found at 3.0-4.0 ppm (Fig. 5, 6). Large peaks at 7.4 ppm in all spectra were derived from water proton remained in the samples.

In the gadolinium-enhanced MR imaging measurements, longitudinal relaxation time of the bulk PBS containing Gd-DTPA reagent showed 0.179 ± 0.06

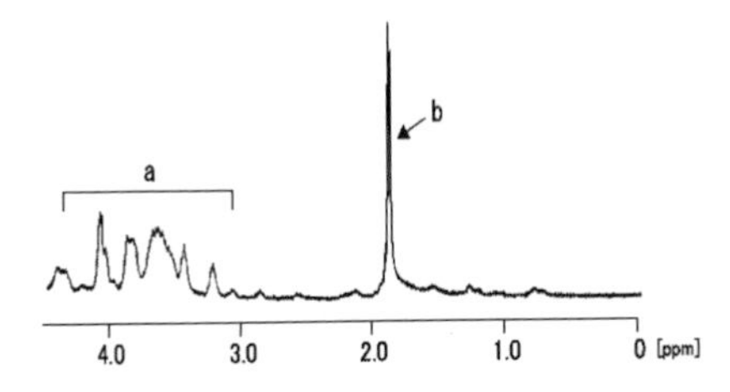

Figure 3. 1H-NMR spectrum of aqueous chondroitin sulfate solution (8 wt.% in D2O). Peaks were found at 2.0 ppm and 3.0-4.5 ppm. [18]

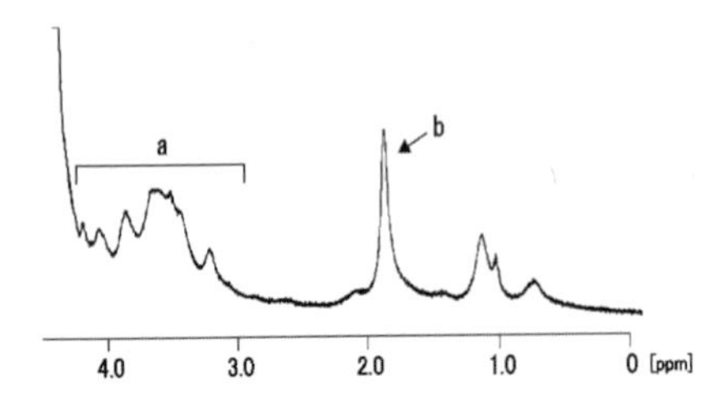

Figure 4. 1H-NMR spectrum of chondrocyte/agarose disk (day42). One narrow peak was found at 2.0 ppm and narrow peaks were found to be imposed by a broad peak at 3.0-4.5 ppm. [18]

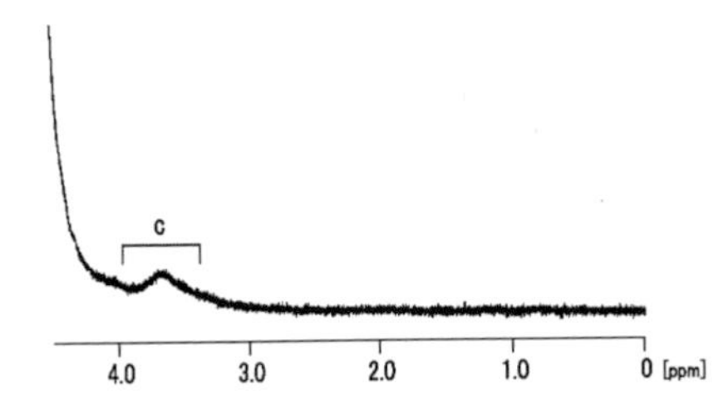

Figure 5. 1H-NMR spectrum of chondrocyte/agarose disk (day2). A broad peak was found at 3.5-4.0 ppm. [18]

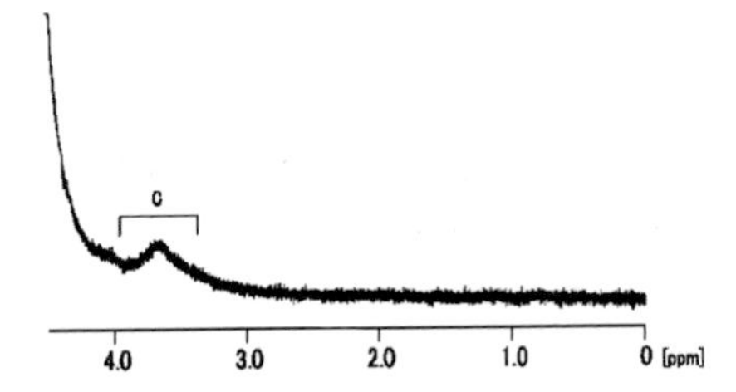

Figure 6. 1H-NMR spectrum of 2% agarose gel (without cell). A broad peak was found at 3.5-4.0 ppm. [18]

seconds in our MRI system. The $T1_{Gd}$ of the cultured specimen increased as a function of tissue maturation (0.197 ± 0.001 to 0.222 ± 0.003 seconds). In the $T1_{Gd}$ maps, the [$Gd\text{-}DTPA^{2-}$] in the specimen decreased, and the boundary between the specimen and the PBS bath became clearer with increased time in culture (Fig. 7). Interestingly, the $T1_{Gd}$ showed higher value in the circumferential area of the disk than in the internal area from day 14 to 28. The FCD calculated from [$Gd\text{-}DTPA^{2-}$] increased according to the time in culture (17.7 ± 1.8 to 40.4 ± 2.2 mM).

As time in culture lengthened, the gross appearance of the cultured disk became increasingly opaque. Typical Safranin O-stained sections of the cultured specimens are shown in Fig. 8. Over the culture time, the chondrocytes in the agarose gel appeared round, similar to the "native" articular cartilage. Figure 8 shows that the chondrocytes synthesized a thin shell of pericellular matrix (~ day10) and expanded the volume of the cartilaginous matrix (~ day28).

The DMMB assay revealed that the sGAG content of the chondrocyte/agarose disks increased as a function of tissue maturation (0.19 ± 0.27 to 13.2 ± 1.9 mg/mL-disk-vol). Finally, the sGAG content of the reconstructed cartilaginous disk reached approximately 20% of the "native" articular cartilage (data not shown). To correlate gadolinium-enhanced MRI and biochemical properties, the sGAG content of the tissue was plotted as a function of the FCD. From the linear regression analysis, the FCD correlated significantly with the sGAG content (r = 0.95, n = 30, P < 0.001) (Fig. 9), and the tissue [Gd-DTPA^{2-}] correlated with the sGAG content by r = 0.83, n = 30, P < 0.001.

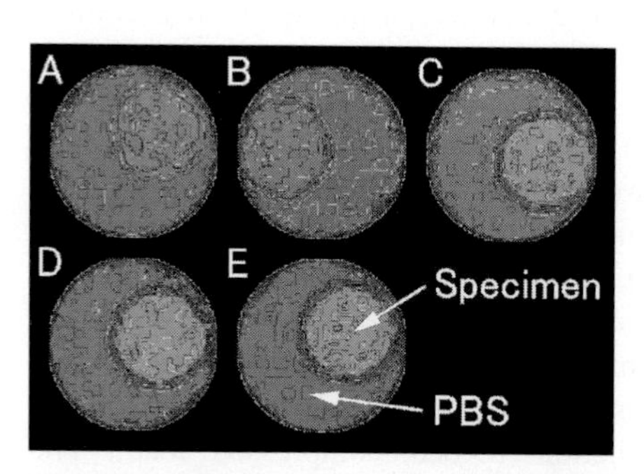

Figure 7. Quantitative water proton T1 maps in the presence of Gd-DTPA^{2-} at day 3 (A), day 7 (B), day 14 (C), day 21 (D), and day 28 (E). [19]

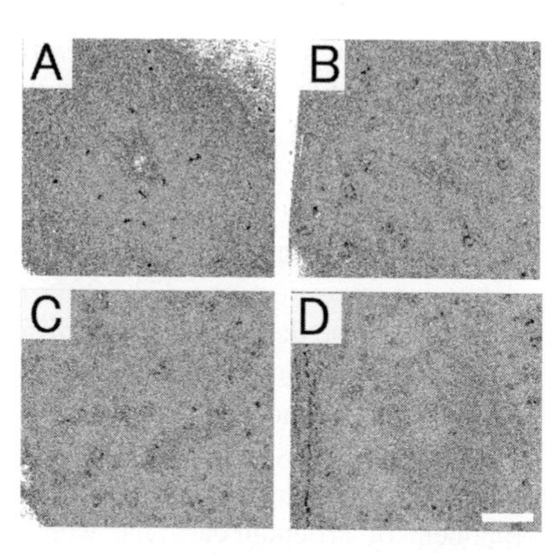

Figure 8. Histological appearance of chondrocyte/agarose disk stained by Safranin O at day 1 (A), day 10 (B), day21 (C), and day 28 (D). [19]

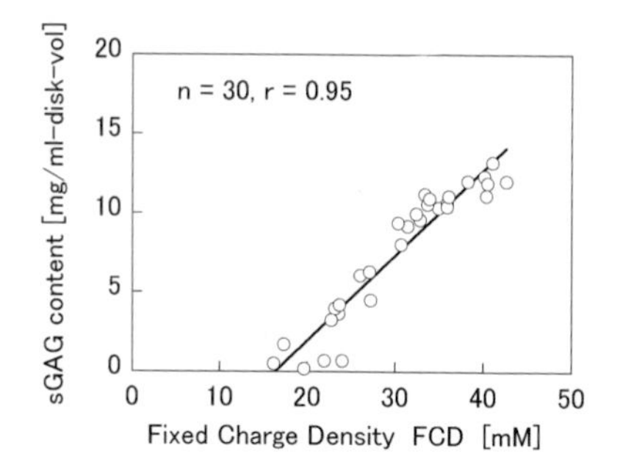

Figure 9. Scatter plots relating the tissue fixed charge density (FCD) to the sulfated glycosaminoglycan (sGAG) content. The correlation between the fixed charge density and the sGAG content can be clearly seen (r = 0.95, n = 30, $P < 0.001$). [19]

4 Discussion

The present study established nondestructive method by NMR spectroscopy to evaluate the molecular structure and noninvasive method by Gd-DTPA enhanced MRI to evaluate the biochemical properties of tissue-engineered cartilage. In addition, we determined the feasibility of these evaluation methods. Similarly to previous investigation [21, 22], we used agarose gel culture to regenerate cartilaginous tissue. The agarose gel is superior in homogeneity and stability in both biomechanical and biochemical properties during the *in vitro* culture. Although the findings from this culture model may not be directly applied to the assessment of regenerated cartilage using porous scaffold material, we adopted the agarose culture method, which enabled more detailed evaluation of the relationship between the biochemical properties and NMR spectra and MRI parameters. In addition, we performed for the first time NMR spectroscopy and Gd-DTPA enhanced MRI measurements on tissue-engineered cartilage consisting of articular chondrocytes.

In this study, we revealed that the chondrocytes in agarose gel synthesized cartilaginous matrix containing sulfated glycosaminoglycan during *in vitro* culture. This finding showed that chondrocytes in the agarose gel were retained in their differentiated phenotype, and reconstructed cartilaginous matrix consisted mainly of collagen and proteoglycan. The findings of our study qualitatively agree with those of previous investigations [21, 22]; however, the values differ slightly. This difference may be derived from the differences of the cultured condition (volume of feed medium, concentration of serum, etc.), the age of the calves, and the region of joints from which chondrocytes were harvested.

In the NMR spectroscopy measurements, we used the chondroitin sulfate solution as a standard for analysis. This is because chondroitin sulfate is a major component of articular cartilage which functions to bear compressive load. The spectrum of the chondroitin sulfate from our experiment showed similar peaks to the spectrum reported by Mucci et al [23]. A peak in 2.0 ppm is derived form the proton of N-acetyl chain in chondroitin sulfate, and peaks in 3.0 – 4.0 ppm are derived from the protons of C–H in chondroitin sulfate (Fig. 10). In the spectrum of chondrocyte-seeded agarose gel cultured for 42 days, peaks were found at almost same position of the standard chondroitin sulfate solution. This result suggests that molecular structure of chondroitin sulfate was reconstructed in the agarose gel. Peaks were also found at 1.0 ppm in the spectrum of the cultured specimen (day42), which might be derived from other components of articular cartilage. A broad peak

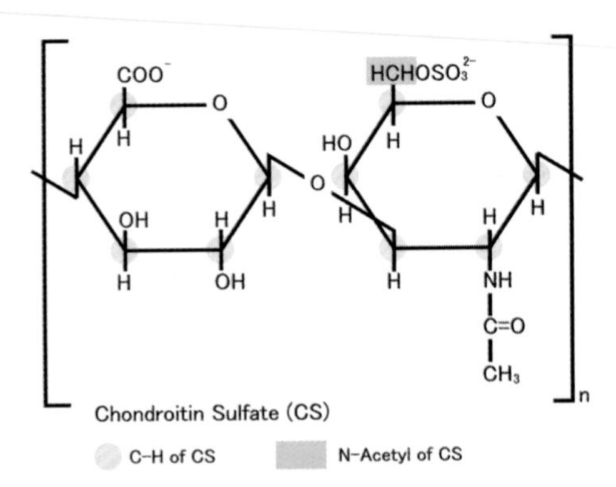

Figure 10. Molecular structures of the subunit of poly-saccharides of proteoglycan in articular cartilage: Chondroitin-6-sulfate. [18]

ranging from 3.0 to 4.0 ppm was found in the spectrum of both the 2% agarose gel and the chondrocyte/agarose culture at day2. This result suggest that the broad peak at 3.0 – 4.0 ppm is derived from protons in agarose molecules. From these results, our nondestructive method using NMR spectroscopy has a potential to evaluate molecular structure of tissue-engineered cartilage. In future study, we will expand this evaluation method to use multiple standard solutions such as hyaluronic acid, keratin sulfate, and collagen.

In the MRI measurements, our results showed that the longitudinal relaxation time $T1_{Gd}$ in the presence of Gd-$DTPA^{2-}$ increased with culture time and that the matrix negative FCD determined by the MRI measurements increased during the *in vitro* culture. These findings are derived from the increasing exclusion of the anionic contrast agent, Gd-$DTPA^{2-}$, by the matrix negative charge of cartilage proteoglycans according to the maturation of tissue. Considering the [Gd-$DTPA^{2-}$] in cartilaginous tissue is relative to the tissue maturation, the FCD measurement in gadolinium-enhanced MRI could be a good predictor of the biochemical properties of regenerated cartilage.

To confirm the relationship between gadolinium-enhanced MRI and the biochemical properties of the regenerated cartilage, we performed linear regression analysis between the FCD and sGAG content of the regenerated cartilage. Our results showed that the FCD determined by gadolinium-enhanced MRI correlated strongly with the sGAG content of regenerated cartilage. This correlation is consistent with previous investigations of FCD measurements of "native" cartilage explants [15].

We note that because our method of evaluating tissue maturity requires a certain level of homogeneity of tissue in the MRI measurement, our methods could be carried on only the tissue-engineered cartilage using gel-like scaffold, not using porous scaffold materials. In addition, to evaluate tissue maturity and development of regenerated cartilage more precisely, other matrix components, such as collagen hyaluronic acid and biomechanical properties must be measured. In future study, we will investigate regenerated cartilage using porous scaffold materials and directly measure other biochemical components and biomechanical properties. We will also determine the relationships between other MRI-derived parameters and material properties to extend our noninvasive method of evaluating regenerated articular cartilage. Although these limitations remain in the evaluation method using gadolinium-enhanced MRI, our findings greatly benefit the assessment of tissue-engineered cartilage using gel-like scaffold materials. Moreover, our findings may be available to assess other regenerated tissue containing negative fixed charge, such as the intervertebral disk.

In conclusion, we evaluate the spectra of tissue-engineered cartilage by 1H-NMR spectroscopy and the change of matrix FCD by Gd-DTPA-enhanced MRI. From the results, we demonstrated that the molecular structure of chondroitin sulfate was regenerated in the engineered cartilage, and that the FCD by the MRI measurements showed a linear correlation with sGAG content of the engineered cartilage. We suggest that our methods using NMR spectroscopy and Gd-DTPA-enhanced MRI can be a useful noninvasive and nondestructive approach to assess the maturity of tissue-engineered cartilage.

Acknowledgments

This research was supported in part by the Special Coordination Funds for Promoting Science and Technology, by a Grant-in-Aid for Young Scientists (B) (No. 18700414) from the Ministry of Education, Science, Sports and Culture of Japan, and by a grant from Nakatani Electric Measuring Technology Association of Japan.

References

1. Mow, V.C., Kuei, S.C., Lai, W.M., Armstrong, C.G., 1980. Biphasic creep and stress relaxation of articular cartilage in compression? Theory and experiments, J. Biomech. Eng. 102, 73-84.
2. Lee, R.C., Frank, E.H., Grodzinsky, A.J., Roylance, D.K., 1981. Oscillatory compressional behavior of articular cartilage and its associated electromechanical properties, J. Biomech. Eng. 103, 280-292.
3. Hunziker, E.B., 1999. Articular cartilage repair: are the intrinsic biological constraints undermining this process insuperable?, Osteoarthritis Cartilage 7, 15-28.
4. Langer, R.S., Vacanti, J.P., 1999. Tissue engineering: the challenges ahead, Sci. Am. 280, 86-89.
5. Wakitani, S., Goto, T., Young, R.G., Mansour, J.M., Goldberg, V.M., Caplan, A.I., 1998. Repair of large full-thickness articular cartilage defects with allograft articular chondrocytes embedded in a collagen gel, Tissue Eng. 4, 429-444.
6. Aoki, H., Tomita. N., Morita. Y., Hattori. K., Harada. Y., Sonobe. M., Wakitani. S., Tamada. Y., 2003. Culture of chondrocytes in fibroin-hydrogel sponge, Biomed. Mater. Eng. 13, 309-316.
7. Chen, G., Sato, T., Ushida, T., Hirochika, R., Tateishi, T., 2003. Redifferentiation of dedifferentiated bovine chondrocytes when cultured *in vitro* in a PLGA-collagen hybrid mesh, FEBS Lett. 542, 95-99.
8. Burgkart, R., Glaser, C., Hyhlik-Durr, A., Englmeier, K.H., Reiser, M., Eckstein, F., 2001. Magnetic resonance imaging-based assessment of cartilage loss in severe osteoarthritis: accuracy, precision, and diagnostic value, Arthritis Rheum. 44, 2072-2077.
9. McCauley, T.R., Disler, D.G., 2001. Magnetic resonance imaging of articular cartilage of the knee. J. Am. Acad. Orthop. Surg. 9, 2-8.

10. Schiller, J., Naji, L., Huster, D., Kaufmann, J., Arnold, K., 2001. 1H and 13C HR-MAS NMR investigations on native and enzymatically digested bovine nasal cartilage, MAGMA 13, 19–27.
11. Schiller, J., Huster, D., Fuchs, B., Naji, L., Kaufmann, J., Arnold, K., 2004. Evaluation of cartilage composition and degradation by high-resolution magic-angle spinning nuclear magnetic resonance, Methods. Mol. Med. 101, 267–285.
12. Potter, K., Butler, J.J., Horton, W.E., Spencer, R.G., 2000. Response of engineered cartilage tissue to biochemical agents as studied by proton magnetic resonance microscopy, Arthritis Rheum. 43, 1580-1590.
13. Gray, M.L., Burstein, D., Xia, Y., 2001. Biochemical (and functional) imaging of articular cartilage, Semin. Musculoskelet. Radiol. 5, 329-343.
14. Shapiro, E.M., Borthakur, A., Kaufman, J.H., Leigh, J.S., Reddy, R., 2001. Water distribution patterns inside bovine articular cartilage as visualized by 1H magnetic resonance imaging, Osteoarthritis Cartilage 9, 533-538.
15. Bashir, A., Gray, M.L., Burstein, D., 1996. Gd-DTPA^{2-} as a measure of cartilage degradation. Magn. Reson. Med. 36, 665-673.
16. Alparslan, L., Minas, T., Winalski, C.S., 2001. Magnetic resonance imaging of autologous chondrocyte implantation, Semin. Ultrasound CT MR 22, 341-351.
17. Miyata, S., Furukawa, K., Ushida, T., Nitta, Y., Tateishi, T., 2004. Static and dynamic mechanical properties of extracellular matrix synthesized by cultured chondrocytes, Mater. Sci. Eng. C 24, 425-429.
18. Miyata, S., Homma, K., Furukawa, K., Ushida, T., Tateishi, T., 2006. Structure assessment of regenerated cartilage using NMR spectroscopy, Trans. JSME C (Japanese) 72, 223-228.
19. Miyata, S., Homma, K., Numano, T., Furukawa, K., Tateishi, T., Ushida, T., 2006. Assessment of fixed charge density in regenerated cartilage by Gd-DTPA-enhanced MR imaging, Magn. Reson. Med. Sci. 5, 73-78.
20. Farndale, R.W., Buttle, D.J., Barrett, A.J., 1986. Improved quantitation and discrimination of sulphated glycosaminoglycans by use of dimethylmethylene blue, Biochim. Biophys. Acta. 883, 173-177.
21. Buschmann, M.D., Gluzband, Y.A., Grodzinsky, A.J., Kimura, J.H., Hunziker, E.B., 1992. Chondrocytes in agarose culture synthesize a mechanically functional extracellular matrix, J. Orthop. Res. 10, 745-758.
22. Mauck, R.L., Soltz, M.A., Wang, C.C., Wong, D.D., Chao, P.H., Valhmu, W.B., Hung, C.T., Ateshian, G.A., 2000. Functional tissue engineering of articular cartilage through dynamic loading of chondrocyte-seeded agarose gels, J. Biomech. Eng. 122, 252-260.
23. Mucci, A., Schenetti, L., Volpi, N., 2000. 1H and 13C nuclear magnetic resonance identification and characterization of components of chondroitin sulfates of various origin, Carbohydrate Polymers 41, 37–45.

SCAFFOLDING TECHNOLOGY FOR CARTILAGE AND OSTEOCHONDRAL TISSUE ENGINEERING

G. CHEN, N. KAWAZOE AND T. TATEISHI

Biomaterials Center, National Institute for Materials Science, 1-1 Namiki, Tsukuba, Ibaraki 305-0044, Japan
E-mail: Guoping.CHEN@nims.go.jp

T. USHIDA

Division of Biomedical Materials and Systems,
Center for Disease Biology and Integrative Medicine, School of Medicine,
The University of Tokyo, 7-3-1 Hongo, Tokyo 113-0033, Japan

Three-dimensional biodegradable porous scaffolds play an important role in cartilage and osteochondral tissue engineering as temporary templates for transplanted cells to guide the new tissue formation. Various scaffolds have been prepared from biodegrabale polymers. Hybridization of biodegradable synthetic polymers and naturally derived polymers has been realized to combine their respective advantages. The hybrid scaffolds facilitate cell adhesion, promote cell proliferation and cartilage tissue formation. Biphasic hybrid scaffold with a stratified two-layer structure for osteochondral tissue engineering has been developed from biodegradable synthetic and naturally derived polymers. The biphasic hybrid scaffold promotes artilage- and bone-like tissues in the respective layers. These recent developments are summarized.

1 Introduction

Once damaged, articular cartilage fails to heal to recover its full functions because of its avascularity and low cellularity. Tissue engineering has been rapidly developed as one of the most promising alternative therapies for articular cartilage defects. Tissue engineering involves the expansion of cells from a small biopsy, followed by the culturing of the cells in temporary three-dimensional scaffolds to form the new cartilage implant. By using the patient's own cells, this approach has the advantages of autografts, but without the problems associated with adequate donor supply. Temporary porous scaffolds play an important role in manipulating cell functions [1-4]. Isolated and expanded cells adhere to the temporary scaffold, proliferate, secrete their own extracellular matrices (ECM) and form a new tissue, replacing the biodegrading scaffold. The porous scaffolds used for cartilage tissue engineering should permit cell adhesion, promote cell proliferation and differentiation, be biocompatible, biodegradable, mechanically strong, and capable of being formed into desired shapes.

A number of three-dimensional porous scaffolds fabricated from various kinds of biodegradable materials have been developed and used for cartilage tissue engineering [5-7]. Especially, polymer materials have received increasing attention and been widely used for cartilage tissue engineering because of easy control over biodegradability and processability. In this review, we will review recent developments of porous polymeric scaffolds for cartilage and osteochondral tissue engineering.

2 Hybrid Porous Scaffolds

There are two kinds of polymer materials: synthetic polymer and naturally derived polymers. The main biodegradable synthetic polymers include polyesters, polyanhydride, polyorthoester, polycaprolactone, polycarbonate, and polyfumarate. The polyesters such as poly (glycolic acid) (PGA), poly (lactic acid) (PLA), and their copolymer of poly (lactic-co-glycolic acid) (PLGA) are most commonly used for tissue engineering. They have gained the approval of the US Food and Drug Administration for certain human clinical use, such as surgical sutures and some implantable devices. PLA undergoes hydrolytic scission to its monomeric form, lactic acid, which is eliminated from the body by incorporation into the tricarboxylic acid cycle. The principal elimination path for lactic acid is respiration, and it is primarily excreted by lungs as CO_2. PGA can be broken down by hydrolysis, nonspecific esterases and carboxypeptidases. The glycolic acid monomer is either excreted in the urine or enters the tricarboxylic acid cycle. The naturally derived polymers include proteins of natural extracellular matrices such as collagen and glycosaminoglycan, alginic acid, chitosan, polypeptides, and etc.

The synthetic and naturally derived polymers have their respective advantages and drawbacks. The biodegradable synthetic polymers can be easily formed into designed shapes with relatively high mechnical strength. However, the scaffold surface of these polymers is relatively hydrophobic, which is not good for cell seeding. On the other hand, naturally derived polymers have specific cell interaction peptides, and their scaffolds have hydrophilic surfaces, which are beneficial to cell seeding and cell attachment. However, naturally derived polymers are mechanically too weak.

The two kinds of polymers have been hybridized to combine their advantages and to avoid their drawbacks by forming microsponges of a naturally derived polymer in the openings of a synthetic polymer skeleton (Fig. 1). The synthetic polymer skeleton allows for easy formation into the desired shapes, and provides the appropriate mechanical strength, while the nested microsponges of naturally derived polymers facilitate cell seeding and cell attachment. Two kinds of such hybrid scaffolds have been reported [8-11]. One is hybrid sponge prepared by introducing collagen microsponges in the pores of a PLGA sponge. The other one is hybrid mesh prepared by forming collagen microsponges in the interstices of PLGA knitted mesh.

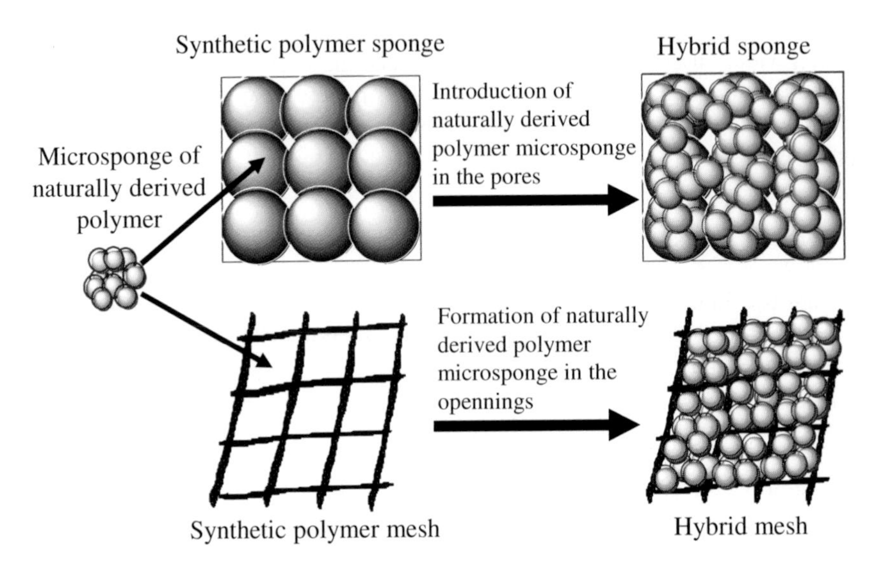

Figure 1. Hybridization of biodegradable synthetic polymers and naturally derived polymers.

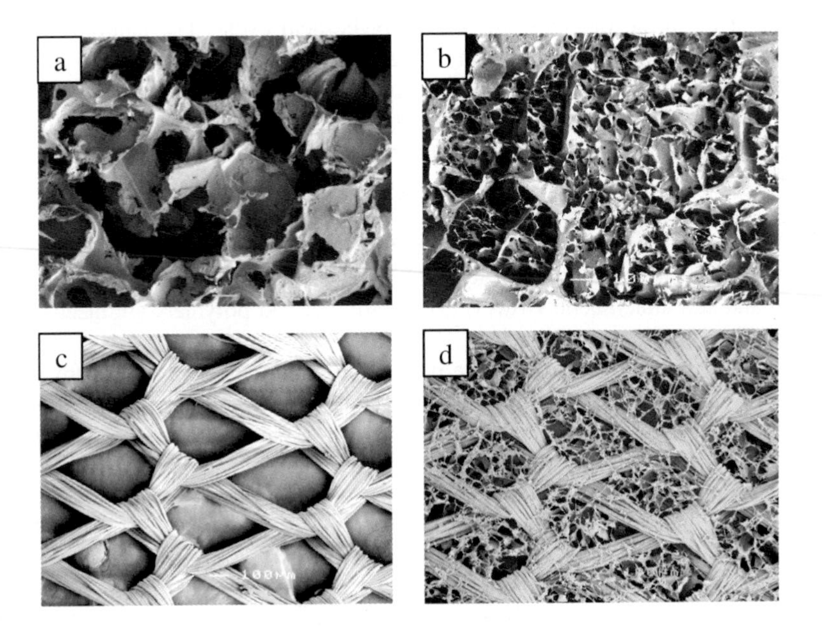

Figure 2. SEM photomicrographs of PLGA sponge (a), PLGA-collagen hybrid sponge (b), PLGA knittd mesh (c) and PLGA-collagen hybrid mesh (d).

The PLGA-collagen hybrid sponge was prepared by immersing a PLGA sponge in a bovine collagen type I acidic solution under a negative pressure, freezing at –80 °C, freeze-drying, and cross-linking with glutaraldehyde vapor. The hybrid structure of the PLGA-collagen hybrid sponge was confirmed by scanning electron microscopy (Fig.2b). Collagen microsponges with interconnected pore structures were formed in the pores of the PLGA sponge. SEM-electron probe microanalysis of elemental nitrogen indicates that microsponges of collagen were formed in the pores of the PLGA sponge and that the pore surfaces were coated with collagen.

The ultimate tensile strength, modulus of elasticity and static stiffness of PLGA-collagen hybrid sponge were higher than those of PLGA and collagen sponges both in dry and wet states. Although the collagen sponge had very low mechanical strength, it was able to reinforce PLGA sponge by forming collagen microsponges in the pores of PLGA sponge. After soaking with HEPES buffer, the mechanical strength of these sponges decreased, especially in the case of collagen sponge.

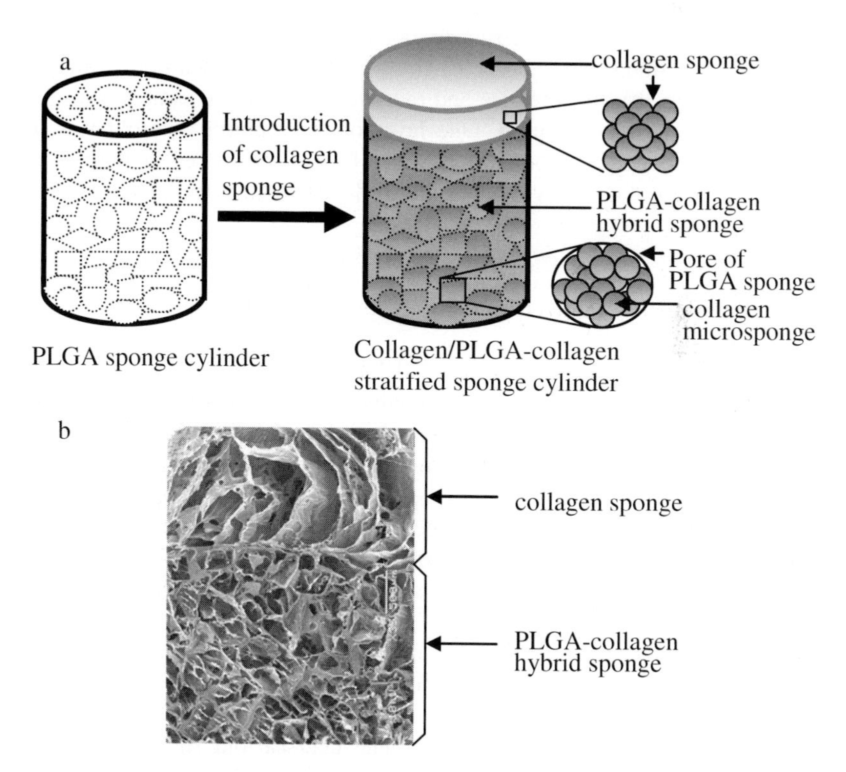

Figure 3. Preparation scheme (a) and SEM photomicrograph (b) of collagen/PLGA-collagen biphasic sponge.

The PLGA-collagen hybrid mesh was prepared by forming collagen microsponges in the openings of a PLGA knitted mesh. SEM observation shows that web-like collagen microsponges were formed in the openings of the synthetic PLGA mesh (Fig.2d). The moduli of elasticity of the hybrid mesh, PLGA mesh and collagen sponge were 35.4±1.4, 35.2±1.0 and 0.020±0.001 MPa, respectively. The hybrid mesh exhibited a significantly higher tensile strength than the collagen sponge alone, similar to the PLGA mesh

3 Biphasic Porous Scaffold

A biphasic scaffold composed of a collagen sponge upper layer and a PLGA-collagen hybrid sponge lower layer can be prepared as shown in Fig.3a. At first, a biodegradable PLGA sponge cylinder is prepared by adding NaCl particulates to a PLGA solution in chloroform and leaching them out of the dried PLGA/NaCl composite. Then, a collagen/PLGA-collagen biphasic sponge cylinder is prepared by introducing collagen sponge into the pores of the PLGA sponge and forming collagen sponge at one side of the PLGA sponge.

SEM observation shows that the PLGA sponge was highly porous with an open pore structure (Fig.3b). Its pore size and morphology were the same as those of the sodium chloride particulates used. One layer of the biphasic scaffold was highly porous collagen sponge. The other layer was a hybrid sponge with collagen sponge formed in the pores of a PLGA sponge. The collagen sponges in the two layers were connected [12].

4 Cartilage Tissue Engineering Using Hybrid Scaffolds

The PLGA-collagen hybrid mesh was used for three-dimesional culture of bovine articular chondrocytes [13]. Subcultured bovine articular chondrocytes were seeded into the PLGA-collagen hybrid mesh, and cultured in vitro in culture media in a 5% CO_2 atmosphere at 37°C. The chondrocytes adhered to the hybrid mesh, proliferated and regenerated cartilaginous matrix filling the void spaces in the hybrid mesh. The web-like collagen microsponge that formed in the openings of the knitted mesh not only prevented the seeded cells going through the composite web, but also increased the specific surface area to provide sufficient surfaces for a spatially even chondrocyte distribution.

After being cultured in vitro for 1 day, the cell/scaffold sheets were used in single form to regenerate thin cartilage implants, or in laminated form to yield thick cartilage implants. The thickness of the implant could be manipulated by changing the number of laminated scaffold sheets. The cell/scaffold sheets could also be rolled up in the shape of a cylinder in which case the thickness of the implant was adjusted by the roll height and its diameter by the rolling number. The round disk-shaped single sheet, 5-sheet and 8 mm-high roll implants were cultured in DMEM for another week and implanted subcutaneously in the dorsum of athymic nude mice. The implants were harvested after 4, 8, and 12 weeks. Gross examination of these grafts showed that all the implants retained their original shapes for all implantation periods and appeared pearly white (Fig.4). The thickness of the engineered cartilage implants of single sheet, 5-sheet and rolled implants were 200 μm, 1 mm and 8 mm respectively.

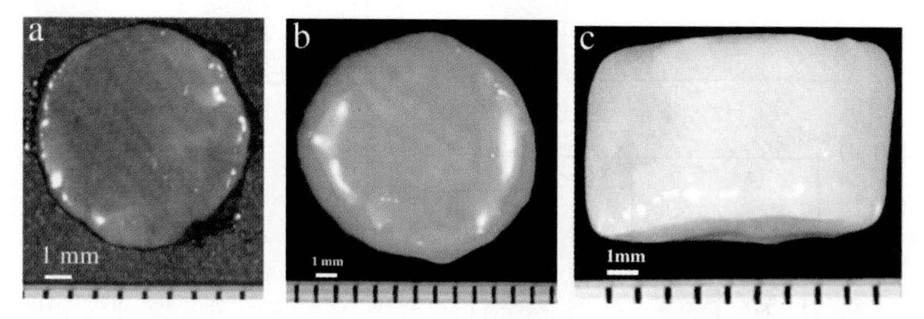

Figure 4. Gross appearance of one-sheet (a), five-sheet (b) and rolled-sheet (c) implant after 12 weeks implantation

Histological examination of these specimens using hematoxylin and eosin stains indicates a uniform spatial cell distribution throughout all the implants both radially and longitudinally. The implant sheets became integrated with each other for the laminated and rolled implants. The chondrocytes in all the implants remained viable, proliferating and secreting extracellular matrix components to form homogeneously compact cartilage tissues. The chondrocytes showed a natural round morphology in all the implants. The bright safranin-O-positive stain indicated that glycosaminoglycans (GAG) were abundant and homogeneously distributed throughout the implants. Toluidine blue staining demonstrated the typical metachromasia of articular cartilage, coinciding with the results of safranin-O staining. Immunohistological staining with an antibody to type II collagen showed a homogeneous extracellular staining for type II collagen. The similarity of the results for the 5-sheet and roll implants to those of the single sheet implant suggests that an increase in the implant thickness from 200 μm to 8 mm does not compromise cell viability, cell uniformity or cellular function.

The mechanical properties of the 5-sheet implant after 12 weeks and bovine native articular cartilage were evaluated by a dynamic compression mechanical test using a viscoelastic spectrometer (Fig.5). The dynamic complex modulus (E*), structural stiffness and phase lag (tanδ) measured at 11 Hz , reached 37.8%, 57.0% and 86.3% of those of native bovine articular cartilage, respectively. These results suggest the formation of articular cartilage. The spatially even distribution of a sufficient number of chondrocytes facilitated the regeneration of articular cartilage. Articular cartilage patches with a thickness ranging from 200 μm to 8 mm were produced by laminating or rolling the hybrid mesh sheets.

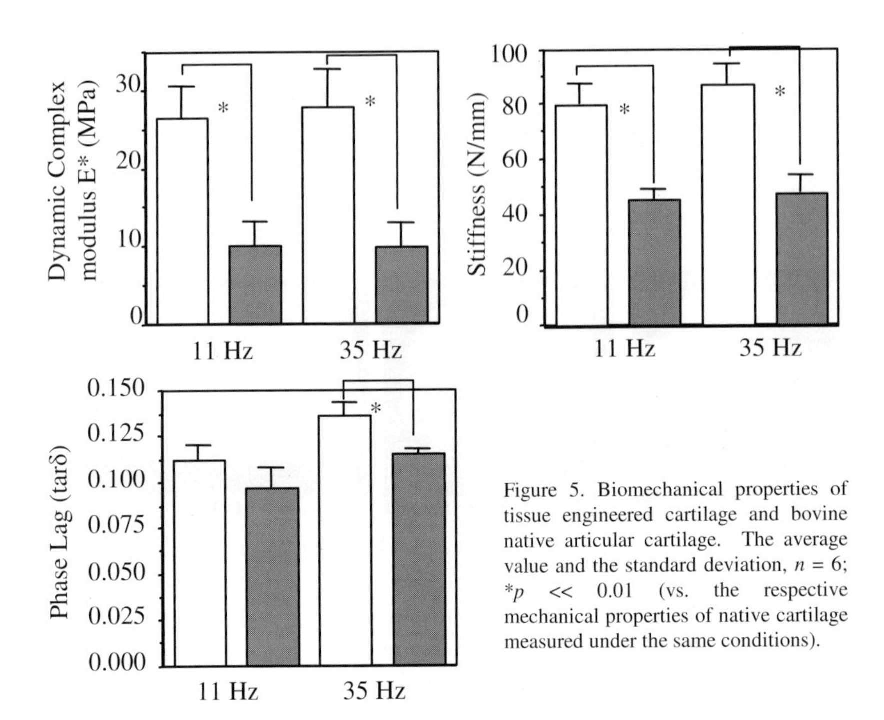

Figure 5. Biomechanical properties of tissue engineered cartilage and bovine native articular cartilage. The average value and the standard deviation, $n = 6$; $*p \ll 0.01$ (vs. the respective mechanical properties of native cartilage measured under the same conditions).

The PLGA-collagen hybrid sponge was also used as the three-dimensional scaffold for tissue engineering of bovine articular cartilage both in vitro and in vivo [14]. In vitro study shows that hybridization with collagen facilitated cell seeding in the sponge and raised seeding efficiency. Chondrocytes adhered on the collagen microsponges of the hybrid sponge, proliferated and secreted extracellular matrices with culture time to fill the space in the sponge. Hematoxylin and eosin staining revealed that most of the chondrocytes after four weeks culture, and most all the cells after cultured for six weeks maintained their phenotypical round morphology. In vivo study by implanting subcutaneously in nude mice demonstrates a more homogeneous tissue was formed in the hybrid sponge than that in the PLGA sponge. The new tissue formed in the hybrid sponge maintained the original shape of the hybrid sponge. The synthetic PLGA sponge serving as a skeleton facilitated easy formation of the hybrid sponges into desired shapes and provided appropriate mechanical strength to define the ultimate shape of engineered tissue. And the incorporated microsponges of collagen facilitated cell seeding, homogeneous cell distribution and favorable environment for cell differentiation.

These results indicate that the PLGA-collagen hybrid mesh and sponge promoted cell adhesion, proliferation and facilitated cartilage tissue formation. They will be useful for cartilage tissue engineering.

5 Osteochondral Tissue Engineering Using Hybrid and Biphasic Scaffolds

Clinical application of tissue-engineered cartilage has the problem of fixation and integration with the surrounding host tissue. A challenge strategy to solve these problems is to engineer an osteochondral tissue that has the same structural and mechanical properties as a native cartilage-bone plug to lead to integration with host cartilage and underlying subchondral bone because mosaicplasty, in which a cylindrical plug of osteochondral tissue from a non-load-bearing region is transplanted to the debrided full-depth defect of the articular cartilage, provides excellent integration of subchondral bone into the host tissue.

Various methods and scaffolds have been developed for osteochondral tissue engineering [15-18]. One strategy is to engineer cartilage and bone layers, separately, then bind the two layers together to construct an osteochondral graft by suture or glue. Schaefer et al reported tissue engineering of osteochondral tissue by suturing together a cartilage construct created by culturing chondrocytes in a biodegradable scaffold and an engineered bone-like construct or a subchondral support. Kreklau et al. engineered a biphasic implant by binding an upper polymer fleece seeded with chondrocytes and a natural coralline material made of calcium carbonate and calcite, a synthetic calcium carbonate by a fibrin-cell solution. Gao et al engineered a cartilage construct by culturing bone-marrow-derived mesenchymal stem cells (MSCs) in hyaluronan derivative sponge with transforming growth factor-beta1 (TGF-beta1) and a bone construct by culturing MSCs in a porous calcium phosphate ceramic scaffold. The two constructs were joined together with fibrin sealant to form a composite osteochondral graft. However, the integration between cartilage and bone by these methods was dependent on the maturity of the two components.

We constructed an osteochondral implant by culturing canine bone marrow stromal cells and articular chondrocytes in PLGA-collagen hybrid mesh (Fig.6) [19]. Canine bone marrow stromal cells were cultured in a PLGA-collagen hybrid mesh in osteogenic medium in vitro and laminated to construct an osteo-layer. Canine articular chondrocytes were cultured in the hybrid mesh in DMEM containing 10% FBS and laminated to construct a chondral layer. The osteo- and chondral layers were sutured together and implanted subcutaneously in nude mice. The original round disc shape of the osteochondral constructs was preserved during the implantation. The osteo- and chondral layers appeared red and glistening white, respectively. Histological examination of the implant specimens indicated that stromal cells and chondrocytes were evenly distributed throughout the scaffold. The laminated meshes were bound together and the two layers had a distinct interface between them. The cells showed a round morphology in the chondral layer and a spindle morphology in the osteo-layer. In the chondral layer, spherical chondrocytes were surrounded by an abundant cartilaginous extracellular matrix. The round morphology and positive stain by safranin O and toluidine blue, together with the expression of genes encoding type II collagen and aggrecan suggested the formation of neocartilage in the chondral layer. Expressions of genes encoding type I collagen and osteocalcin were detected in the osteochondral implant. These results indicate the formation of osteochondral-like tissue, and the hybrid mesh and lamination method may be useful for osteochondral tissue engineering.

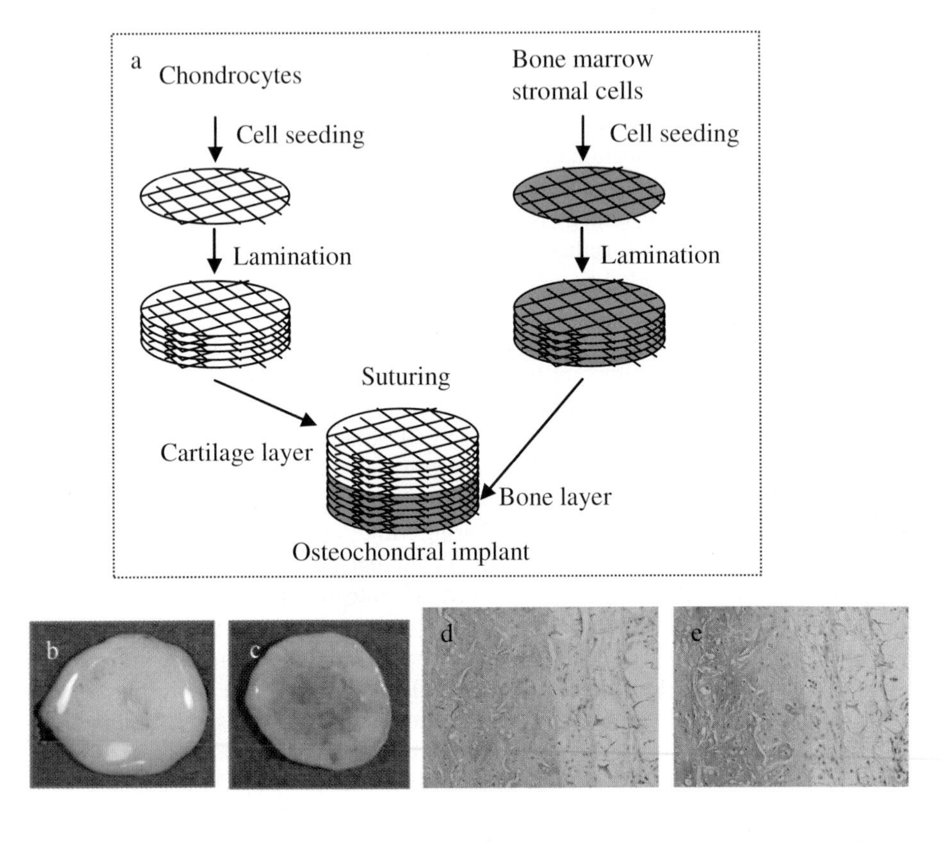

Figure 6. The lamination scheme of PLGA-collagen hybrid meshes seeded with bone marrow stromal cells and chondrocytes (a), gross appearance of the osteochondral tissue (cartilage side (b) and bone side (c)), hematoxylin / eosin (d) and safranin O/fast green (e) staining of the osteochondral tissue after 9 weeks implantation.

Another strategy is to use a biphasic scaffold [20, 21]. Sherwood et al reported a heterogeneous osteochondral scaffold using the TheriForm three-dimensional printing process. The upper, cartilage region was 90% porous and composed of D,L-PLGA/L-PLA; the lower, cloverleaf-shaped bone portion was 55% porous and consisted of a L-PLGA/TCP composite. Schek et al developed a biphasic scaffold composed of poly-L-lactic acid/hydroxyapatite using image-based design (IBD) and solid free-form (SFF) fabrication. The biphasic scaffolds were differentially seeded with fibroblasts transduced with an adenovirus expressing bone morphogenetic protein 7 (BMP-7) in the ceramic phase and fully differentiated chondrocytes in the polymeric phase.

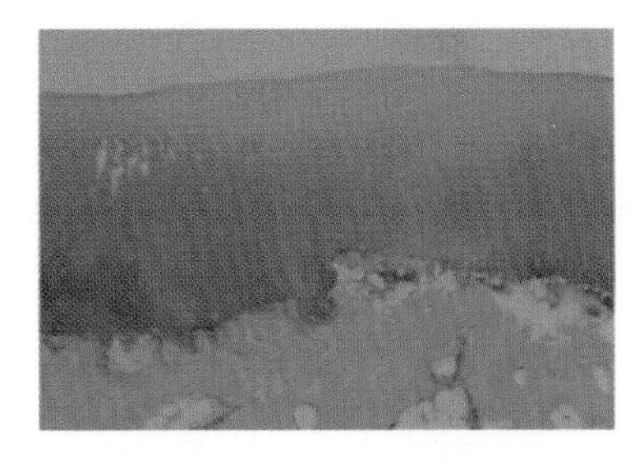

Figure 7. Safranin O/fast green staining of the 4-month implant of collagen/PLGA-collagen biphasic sponge with MSCs.

We applied the collagen/PLGA-collagen biphasic scaffold for oesteochondral tissue engineering [12]. Canine bone-marrow-derived mesenchymal stem cells (MSCs) were isolated from bone marrow aspirate of a one-year-old beagle and subcultured in DMEM containing 10% FBS. The subcultured MSCs were seeded into the collagen/PLGA-collagen biphasic sponge scaffold, cultured in vitro for one week, and transplanted into the knee of the same beagle. The implants were harvested after 3 months. Gross appearance showed that the defect treated with cells/scaffold presented a smoother surface and better integration with the surrounding tissue than did the scaffold without cells. Histological examination of these specimens using hematoxylin and eosin, and safranin O/fast green stains indicated that cartilage-like and underlying bone-like tissues were regenerated 4 months after implantation (Fig.7). The cartilage-like tissue was stained intensively to safranin O and well integrated with the surrounding tissue. However, the defect implanted with scaffold without the cells did not show any evidence of hyaline cartilage regeneration. The collagen/PLGA-collagen biphasic scaffold will be a useful three-dimensional scaffold for osteochodral tissue engineering.

6 Conclusions

Various biodegradable porous scaffolds have been developed by hybridizing and biphasic preparation techniques. The hybrid scaffolds of synthetic polymers and naturally derived polymers combine the advantages of the two kinds of biodegradable polymers. The skeleton of synthetic polymers defines the gross shape and size of the scaffolds and supports the forming tissue during the initial stages, while the incorporated microsponge of naturally derived polymers facilitates cell seeding and destribution. The biophasic porous scaffold has a stratified two-layer structure in which the upper layer was collagen sponge and the lower layer was a hybrid sponge of synthetic PLGA and naturally derived collagen. The biphasic scaffold seeded with MSCs facilitates the integration with the surrounding tissue and promotes osteochondral tissue regeneration when implanted in the osteochondral defect of the knee of a beagle. The hybrid and biphasic porous scaffolds will be useful for cartilage and osteochondral tissue engineering.

Acknowledgments

This work was supported by the Ministry of Education, Culture, Sports, Science and Technology of Japan.

References

1. Sims, C.D., Butler, P.E.M. Cao, Y.L., Casanova, R., Randolph, M.A., Black, A., Vacanti, C.A. Yaremchuk, M.J., 1998. Tissue engineered neocartilage using plasma derived polymer substrates and chondrocytes, Plast. Reconstr. Surg. 101, 1580-1585.
2. Rotter, N. , Aigner, J., Naumann, A., Planck, H., Hammer, C., Burmester, G., Sittinger, M., 1998. Cartilage reconstruction in head and neck surgery: Comparison of resorbable polymer scaffolds for tissue engineering of human septal cartilage, J. Biomed. Mater. Res. 42, 347-356.
3. Chen, G., Ushida, T., Tatsuya, T, 2002. Scaffold design for tissue engineering, Macromole. Biosci. 2, 67-77.
4. Chen, G., Ushida, T., Tatsuya, T, 2001. Development of biodegradable porous scaffolds for tissue engineering, Mater. Sci. Eng. C 17, 63-69.
5. Sherwood, J.K., Riley, S.L., Palazzolo, R., Brown, S.C., Monkhouse, D.C., Coates, M., Griffith, L.G., Landeen, L.K., Ratcliffe, A., 2002. A three-dimensional osteochondral composite scaffold for articular cartilage repair, Biomaterials 23, 4739-4751.
6. Isogai, N., Landis, W., Kim, T.H., Gerstenfeld, L.C., Upton, J., Vacanti, J.P., 1999. Formation of phalanges and small joints by tissue-engineering, J. Bone Joint Surg. Am. 81, 306-316.
7. Peter, S.J., Miller, M.J., Yasko, A.W., Yaszemski M.J., Mikos, A.G., 1998. Polymer concepts in tissue engineering, J. Biomed. Mater. Res. 43, 422-427.
8. Chen, G., Ushida, T., Tatsuya, T, 1999. Fabrication of PLGA-collagen hybrid sponge, Chem. Lett. 28, 561-562.
9. Chen, G., Ushida, T., Tatsuya, T, 2000. Hybrid biomaterials for tissue engineering: a preparative method of PLA or PLGA-collagen hybrid sponge, Adv. Mater. 12, 455-457.
10. Chen, G., Ushida, T., Tatsuya, T, 2000. A biodegradable hybrid sponge nested with collagen microsponges, J. Biomed. Mater. Res. 51, 273-279.

11. Chen, G., Ushida, T., Tatsuya, T, 2000. A hybrid network of synthetic polymer mesh and collagen sponge, J. Chem. Soc. Chem. Comm. 16, 1505-1506.
12. Chen, G., Sato, T., Tanaka, T., Tatsuya, T, 2006. Preparation of a biphasic scaffold for osteochondral tissue engineering, Mater. Sci. Eng. C 26, 118-123.
13. Chen, G., Sato, T., Ushida, T., Hirochika, R., Shirasaki , R., Ochiai, N., Tetsuya, T, 2006. The use of a novel PLGA fiber/collagen composite web as a scaffold for engineering of articular cartilage tissue with adjustable thickness, J. Biomed. Mater. Res. 67A, 1170-1180.
14. Chen, G., Sato, T., Ushida, T., Ochiai, N., Tetsuya, T, 2004. Tissue engineering of cartilage using a hybrid scaffold of synthetic polymer and collagen, Tissue Engineering, 10, 323-330.
15. Schaefer, D., Martin, I., Jundt, G., Seidel, I., Heberer, M., Grodzinsky, A., Bergin, I., Vunjak-Novakovic, G., Freed, L.E., 2002. Tissue-engineered composites for the repair of large osteochondral defects, Arthritis Rheum 46, 2524-2534.
16. Gao, J., Dennis, J.E., Solchaga, L.A., Awadallah, A.S., Goldberg, V.M., Caplan, A.I., 2001. Tissue-engineered fabrication of an osteochondral composite graft using rat bone marrow-derived mesenchymal stem cells, Tissue Eng. 7, 363-371.
17. Schaefer, D., Martin, I., Shastri, P., Padera, R.F., Langer, R., Freed, L.E., Vunjak-Novakovic, G., 2000. In vitro generation of osteochondral composites, Biomaterials 21, 2599-2606.
18. Kreklau, B., Sittinger, M., Mensing, M.B., Voigt, C., Berger, G., Burmester, G.R., Rahmanzadeh, R., Gross, U., 1999. Tissue engineering of biphasic joint cartilage transplants, Biomaterials 20, 1743-1749.
19. Chen, G., Tanaka, T., Tatsuya, T, 2006. Osteochondral tissue engineering using a PLGA-collagen hybrid mesh, Mater. Sci. Eng. C 26, 124-129.
20. Sherwood, J.K., Riley, S.L., Palazzolo, R., Brown, S.C., Monkhouse, D.C., Coates, M., Griffith, L.G., Landeen, L.K., Ratcliffe, A., 2002. A three-dimensional osteochondral composite scaffold for articular cartilage repair, Biomaterials 23, 4739-4751.
21. Schek, R.M., Taboas, J.M., Segvich, S.J. Hollister, S.J., Krebsbach, P.H., 2004. Engineered osteochondral grafts using biphasic composite solid free-form fabricated scaffolds, Tissue Eng. 10, 1376-1385.

IV. COMPUTATIONAL BIOMECHANICS

MRI MEASUREMENTS AND CFD ANALYSIS OF HEMODYNAMICS IN THE AORTA AND THE LEFT VENTRICLE

M. NAKAMURA AND S. WADA

Department of Mechanical Science and Bioengineering, Osaka University,
1-3 Machikaneyama, Toyonaka 560-8531, Japan
E-mail: masanori@me.es.osaka-u.ac.jp

S. YOKOSAWA

HItachi Central Research Laboratory,

1-280, Higashi-Koigakubo, Kokubunji-shi, Tokyo 185-8601, Japan

T. YAMAGUCHI

Department of Bioengineering and Robotics, Tohoku University,
6-6-01 Aoba-yama, Sendai 980-8579, Japan

Flow in a human aorta has a complex nature as being affected by its non-planar configuration, branches and inflow dynamics stemmed from the left ventricle. The aim of the present study is to investigate hemodynamics in the human aorta and the left ventricle by combining magnetic resonance imaging (MRI) measurements and computational fluid dynamics (CFD) simulations, and to see what geometric and flow conditions are of importance to re-produce hemodynamics observed in vivo. MRI was used to define the geometry of a human aorta, whereby aorta models with/without three branches and taper were constructed. The numerical left ventricle was developed and attached to the aortic root to create an integrated model of the left ventricle and the aorta. Cine phase-contrast MRI was used to acquire 3D time-resolved velocities at the inlet and outlets of the aorta. A flow simulation with an integrated model of the aorta and the left ventricle successfully demonstrated a series of flow events in the ventricle and aorta during a cardiac cycle. The flow of ejected blood through the open aortic valve had markedly skewed velocity profiles as observed in *in vivo* with PC-MRI. A comparison of the wall shear stress distribution from the integrated analysis to that obtained in flow calculations for an isolated aorta model with Poiseuille flow and flat inlet conditions showed that the effect of intraventricular flow on the WSS persisted to the top of the aortic arch. A flow simulation with the aorta model with/without tapering and branches commonly demonstrated the development of a helical flow. However, the presence of the branches provoked some flows going upward in the aortic arch, inducing slight disturbances to secondary flows in the descending aorta. The presence of the branches also affected WSS and OSI distribution, although the affected regions were very limited. These results addressed the importance of defining of inflow conditions and modeling of aortic branches for a detailed analysis of the aortic hemodynamics, charting future directions of combined MRI-CFD flow analysis.

1 Introduction

Hemodynamics is deeply related to blood coagulation and thromboembolism [1] endothelial cell structure and function [2], and the uptake and accumulation of molecules on the arterial wall [3]. A great number of attempts have been carried out to relate hemodynamic factors, such as the wall shear stress (WSS) and its derivatives in both time and space, to atherosclerotic lesions and aneurysms [4].

Aorta is frequently involved in atherosclerotic lesions. The damage caused to the aortic wall by atherosclerosis can result in aneurysm formation. Thus, studies to relate the hemodynamics factors to the occurrence of those vascular diseases. Most of these studies, however, idealized an inflow condition; [5, 6], although the inflow dynamics is complex because of the left ventricular flow ejection. In addition, they sometimes adopted a simplified geometry of the aorta. The most frequent and apparent simplification was to ignore the branching and tapering of the aorta to eliminate the difficulty in defining computational grids [5–7].

We have investigated the hemodynamics in the human aorta and the left ventricle by using magnetic resonance imaging (MRI) measurements and computational fluid dynamics (CFD) simulations [8, 9]. This paper deals with two major issues in modeling of the aortic blood flow; (1) the importance of modeling of the aortic inflow and (2) the influence exerted by simplifying the aortic geometry branching and tapering of the aorta. MRI was used to define the geometry of a human aorta, from which aorta models with or without three branches and taper were constructed. Cine phase-contrast MRI was used to acquire 3-D time-resolved velocities at various sections of the aorta and the left ventricle.

2 Methods

2.1 Measurement of the aortic geometry and flow using MRI

Two-dimensional cine phase-contrast MRI with a 1.5-T MR system (Signa Infinity EchoSpeed with the Excite option; General Electric, Fairfield, CT, USA) was used to obtain a magnitude image to provide anatomical information and phase images for attaining flow velocities in three orthogonal directions. The MR parameters were determined using standard settings for clinical examinations: repetition time (TR) 33 ms, echo time (TE) 5.4–6.0 ms, velocity encoding range (VENC) 150 cm/s, flip angle 30°, slice thickness 5 mm, matrix 192 × 192, and field of view (FOV) 32 × 24 cm.

The subject was an adult male volunteer with no history of cardiac disease. Written informed consent was obtained from the volunteer with the approval of the Human Study Committee of Tohoku University and Hamamatsu University School of Medicine. All measurements were taken while the subject was holding his breath after a maximum expiration. The coronal and transverse planes of his chest were scanned to define the aorta geometry. Velocities were measured at the middle plane of the left ventricle, the entrance of the aorta, three aortic branches, and the descending aorta. A series of 30 equidistant images per heartbeat interval were acquired under electrocardiogram (ECG) synchronization. The trigger pulse was obtained from the ECG R-wave.

2.2 Aorta models

Three computational models of the aorta were prepared: a model without branches or taper, a model with branches but no taper, and a model with branches and taper. The geometry of the aorta was determined from the MR images. Given the transverse and coronal planes of the MR images, the centerline of the main trunk of the aorta was approximated by taking several points that were considered to be on the centerline. For all models, we assumed that the cross sections perpendicular to the centerline of the aorta were circular. The diameters of the cross sections were set to 2.6 cm for the ascending aorta, 1.0 cm for the brachiocephalic artery, and 0.7 cm for the left carotid and left subclavian arteries. Care was taken to insure that no sharp corners occurred at the connecting junctions between the aorta and branches. The tapering applied to the aortic arch was estimated by fitting the exponential equation

$$A(x) = A_0 \exp(Cx) \tag{1}$$

to the data set for the cross-sectional area of the aortic arch obtained from the MRI. Here, x is the distance from the aortic inlet along the centerline, A is the cross-sectional area of the aortic arch, A_0 is the cross-sectional area at the aortic inlet, and C is a constant. Using the least squares method, we obtained A_0 = 6.75 cm^2 and C = –0.037. The diameter from the exit of the aortic arch to the descending aorta was set to a constant value of 2.28 cm.

2.3 Left ventricle model

The geometry of the left ventricle model was also based on clinical data [10]. For simplicity, it was assumed to be symmetric with respect to a long-axis plane. The mitral and aortic valves were both circular with radii of 1.3 cm. In modeling the deformation of the left ventricle, it was assumed that the ventricular surface moved inwards or outwards in its normal direction independent of the internal blood pressure and other hemodynamic forces such as wall shear stress. At every time step, the moving velocity of the ventricular wall v_w was calculated such that an increment in the volume $V(t)$ due to the wall movement was equal to a prescribed volume change at time t. Mathematically, it was expressed by

$$\frac{\mathrm{d}V(t)}{\mathrm{d}t} = \iint_S v_w \mathbf{e} \cdot \mathbf{n} \mathrm{d}S , \tag{2}$$

where $\mathbf{e}$ is a unit vector in the direction of the wall movement, dS is the element of the left ventricle model surface area, and n denotes the components of a unit vector normal to the surface S. For further simplicity, the moving velocity of the ventricular wall v_w was factored into

$$v_w = v_a(t)W , \tag{3}$$

where $v_a(t)$ is the velocity at the ventricular apex and the weighting function W. In this simulation, W was formulated such that the wall movement increased from the base to the apex, while the walls did not move at the mitral and aortic valves to maintain the shape of the valves [11]. The mitral valve was modeled as a circular core with no leaflets that has changes its size as a function of a rate of the ventricular volume change [12]. The aortic valve was also modeled as a circular core, although it opened or closed instantaneously. The temporal variation in the left ventricle volume was based on modifications of MRI-derived data [13]. The lengths of systole and diastole were 0.413 and 0.337 s, respectively. The stroke volume was 70 ml.

2.4 *Blood flow model*

Blood was treated as an incompressible Newtonian fluid with a density of 1.03×10^3 kg/m^3 and a viscosity of 4.0×10^{-3} Pa·s. Computations were performed using the commercial CFD program SCRYU ver. 2.11 (Software Cradle, Osaka, Japan), and ANSYS-FLORTRAN ver. 7.1 (Cybernet Systems, Co. Ltd., Japan) to solve the laminar flow described by the Navier–Stokes and continuity equations:

$$\frac{\partial \mathbf{U}}{\partial t} + (\mathbf{U} \cdot \nabla)\mathbf{U} = -\frac{1}{\rho}\nabla P + \nu \nabla^2 \mathbf{U} \tag{4}$$

$$(\nabla \cdot \mathbf{U}) = 0 , \tag{5}$$

where **U** is the 3-D velocity vector, P is the pressure, and ρ and ν are the density and kinematic viscosity of blood, respectively.

2.5 *Simulation condition and procedure*

Two simulations are performed. One is a combined simulation of the left ventricular flow and the aortic flow. The simulation was commenced from diastole during which only the left ventricular flow was calculated with the boundary conditions of zero pressure at the open part of the valve, zero velocity at the closed part of the valve, and the moving velocity of the wall at other superficial nodes of the left ventricle model. At the onset of systole, the ventricle was attached to the root of the aorta and the aortic valve was opened. The systolic flow was calculated under the condition of zero pressure at the outlet of the descending aorta, the moving velocity at the ventricular wall, zero velocity at the aorta and the mitral valve. In this case, the aorta model without branches and taper was used.

The other is a flow simulation in the aorta models with/without tapering and branches described in 2.2. At the inlet, all three components of the velocity data measured with MRI were imposed after mapping the measured MR data onto the computational grid inlet using a coordinate transformation. The flow velocity at a time instant that was not measured was interpolated using a cubic spline function. Parabolic flow was applied at the outlet of the aortic branches. The distribution ratio of flow to the branches and descending aorta was determined from the MR data and fixed throughout the cardiac cycle. It was set such that 15% of the total aortic inflow went to the brachiocephalic artery, 10% to the left carotid artery, and 8% to the left subclavian artery. A nonslip condition was applied at the wall and a traction-free condition was imposed at the end of the descending aorta.

2.6 Hemodynamics factors

The WSS and oscillatory shear index (OSI) [14] were evaluated as hemodynamics factors that might act on the arterial wall to cause vascular diseases. The WSS is calculated simply by multiplying a velocity gradient parallel to the wall with a viscosity. The OSI is defined by

$$\mathrm{OSI} = \frac{1}{2}\left(1 - \left|\int_0^T \boldsymbol{\tau}_w dt\right| \Big/ \int_0^T \left|\boldsymbol{\tau}_w\right| dt\right). \tag{6}$$

3 Results

3.1 Importance of the inflow condition at the aorta

3.1.1 Flow simulation with an integrated model of the left ventricular and the aorta

Figure 1 shows simulated flow patterns, expressed with particle tracking, at (a) mid-diastole, (b) early systole and (c) mid systole and (d) end diastole in the integrated model of the left ventricle and the aorta. The images are viewed from the left of the body with the septum (anterior) side towards the left. With the onset of the left ventricular expansion, blood flowed into the ventricular cavity. From the mid-diastole, fluid elements under the aortic valve started to recirculate clockwise when viewed from its left. As a consequence, a vortex arose under the aortic valve, redirecting blood inflow that headed straight to the apex smoothly towards the outflow tract (Fig. 1a). When systole began, blood was ejected into the aorta (Fig. 1b). Until mid-systole, the blood in the aorta was directed axially, along the curve of the aortic arch (Fig. 1c). As systole progressed further, a helical flow developed within the aorta. By the end of systole, secondary helical flows dominated the entire aorta.

In Fig. 2, contour plots of the axial flow at the aortic valve plane at the same moments as Fig. 1b, 1c and 1d are illustrated with vector plots of secondary flow. The image was viewed from the downstream side with the ventricular septum towards the up and with MV towards the bottom. In early systole, velocity profile was fairly flat (Fig. 2a). In mid systole, velocity profile was skewed to the MV side and a pair of swirls developed across the symmetric plane of the left ventricle (Fig. 2b). With the swirls, high velocity portion of the flow moved to the septum side,

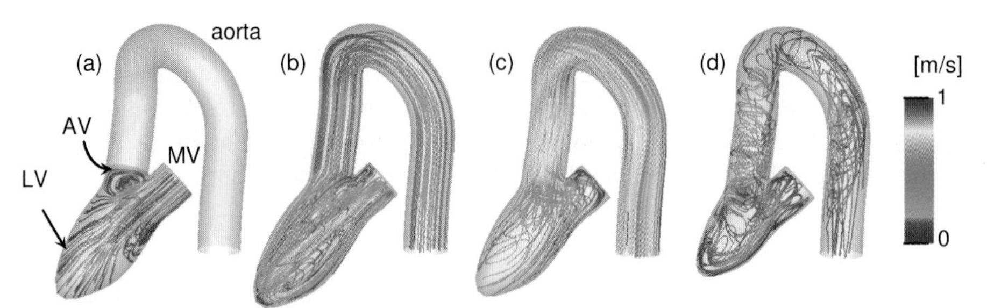

Figure 1. Streamlines of blood flow in (a) mid diastole, (b) early, (c) mid, (d), end systole. Images are viewed from the left of the body with its septum (anterior) side towards the left. LV: left ventricle, AV: aortic valve. MV: mitral valve.

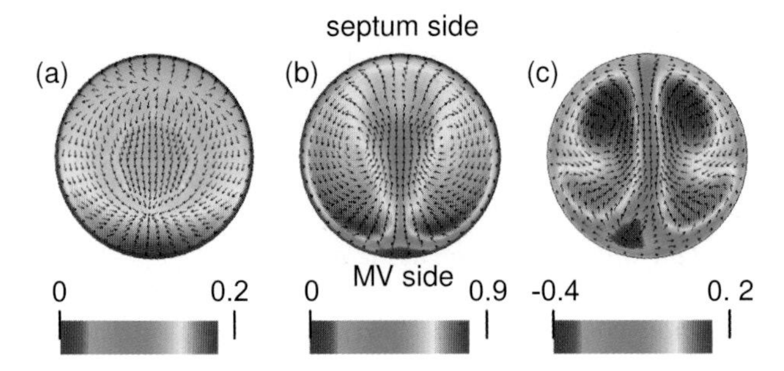

Figure 2. Vector and counter plots of the flow at the plane at the aortic valve constructed from the simulation results. (a) early, (b) mid, (c), end systole.

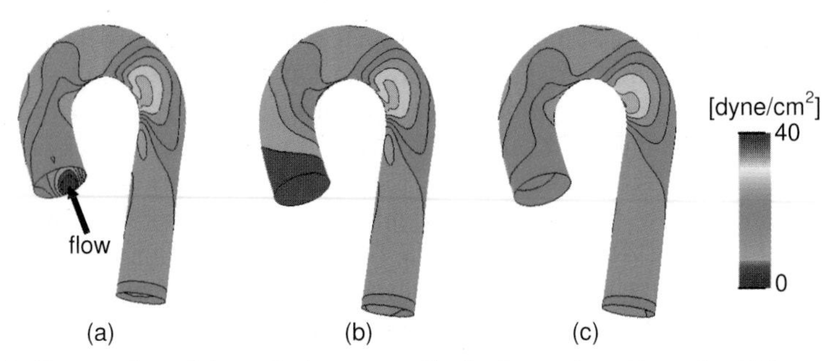

Figure 3. Contour plots of the wall shear stress distribution at the mid-systole. (a) integrated model, (b) Poiseuille flow and (c) flat flow.

and flow through AV became faster close to the ventricular septum than beside MV (Fig. 2c).

The WSS distribution at the aorta obtained with the integrated simulation is compared with that obtained with simulations where the aorta model was isolated from the left ventricle and either Poiseuille or flat velocity profile was given as an inlet boundary condition. A temporal variation of a flow rate given in the simulation

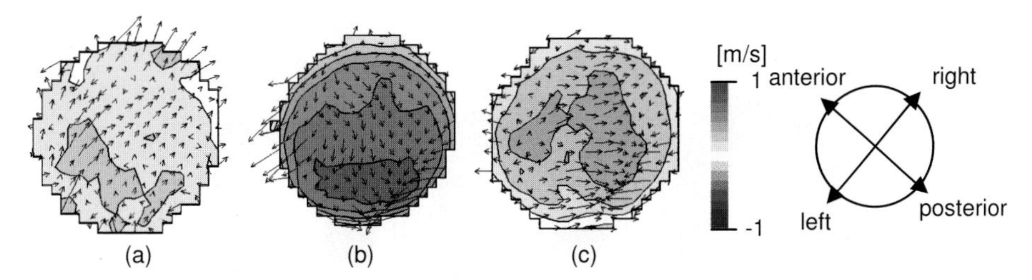

Figure 4. Contour plots of the axial flow and vector plots of the secondary flows just above the aortic valve measured with PC-MRI. (a) early, (b) mid, (c), end systole. Directions of images are shown on the left. A positive velocity indicates a flow going upward.

of the isolated aorta model was the same as the volume change rate of the left ventricle in the integrated simulation. Figure 3 illustrates contour plots of the WSS distribution at the mid-diastole. There was distinct variation in the pattern of the WSS distribution in the ascending aorta for simulations with different inflow conditions. Conversely, beyond the top of the aortic arch, the WSS distribution was similar for all simulations.

3.1.2 Measurement of the flow dynamics just above the aortic annulus

Cross-sectional velocity profiles at the plane 1cm distal to the aortic annulus in early, mid and end systole measured with PC-MRI are presented in Fig. 4. The section is viewed from the downstream side, and a positive flow velocity indicates that the flow travels upward to the head. They demonstrate a non-flat velocity profile throughout a cardiac cycle. In early systole, the flow velocity was relatively high near the anterior side of the vessel (Fig. 4a). As systole progressed, the velocity at the central part increased, resulting in an approximately axisymmetric velocity profile at mid-systole (Fig. 4b). Between middle and late systole, the velocity profile became slightly skewed, with faster flow developing towards the right side of the vessel (Fig. 4c). In late systole, a regurgitating flow was identified at the left side of the vessel. During diastole, the flow velocity was almost zero and contained only small fluctuations.

3.2 Hemodynamics in the aorta models with/without tapering and branches

Figure 5 shows streamlines of flow in (a) the aorta model without the branches and taper, (b) the model with branches but no taper and (c) the model with branches and taper. The flow patterns in the aorta were qualitatively similar for all cases.

Figure 6 compares a spatial distribution of the contour plots of the WSS at the peak of systole. The WSS distribution at the main trunk of the aorta was qualitatively similar. In all models, a relatively high WSS was found at the inner curvature of the proximal (the other side of this figure) and distal ends of the aortic arch, although the WSS for the model with the branches and no taper was slightly lower. For the models with branches, a high WSS occurred at the roots of the

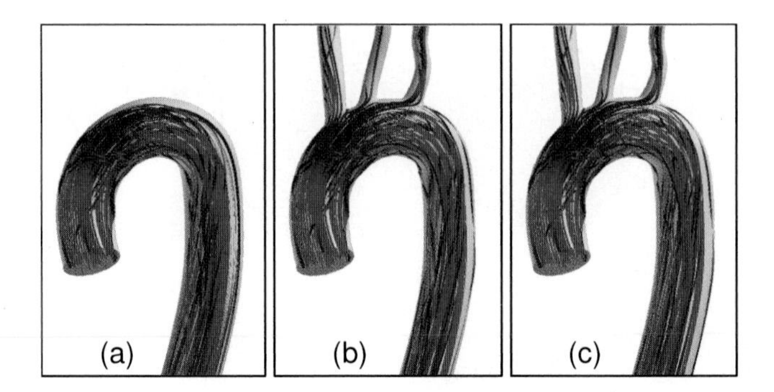

Figure 5. Flow patterns at the peak systole. (a) the model without the branches and taper, (b) the model with branches but no taper and (c) the model with branches and taper.

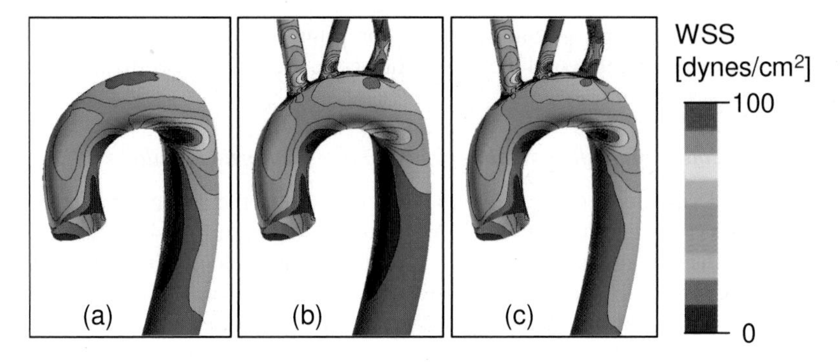

Figure 6. WSS distribution at the peak systole. (a) the model without the branches and taper, (b) the model with branches but no taper and (c) the model with branches and taper.

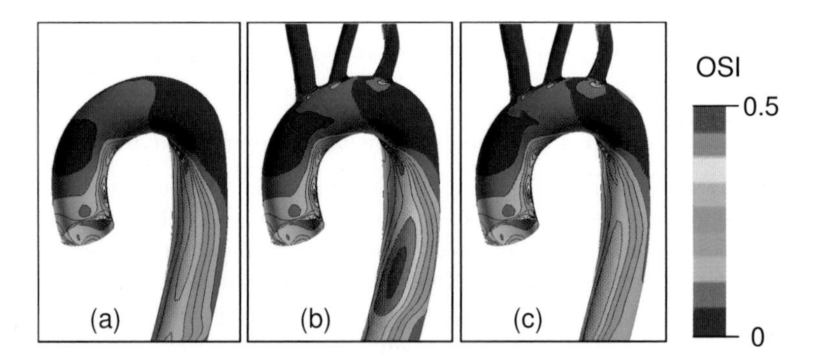

Figure 7. OSI distribution at the peak systole. (a) the model without the branches and taper, (b) the model with branches but no taper and (c) the model with branches and taper.

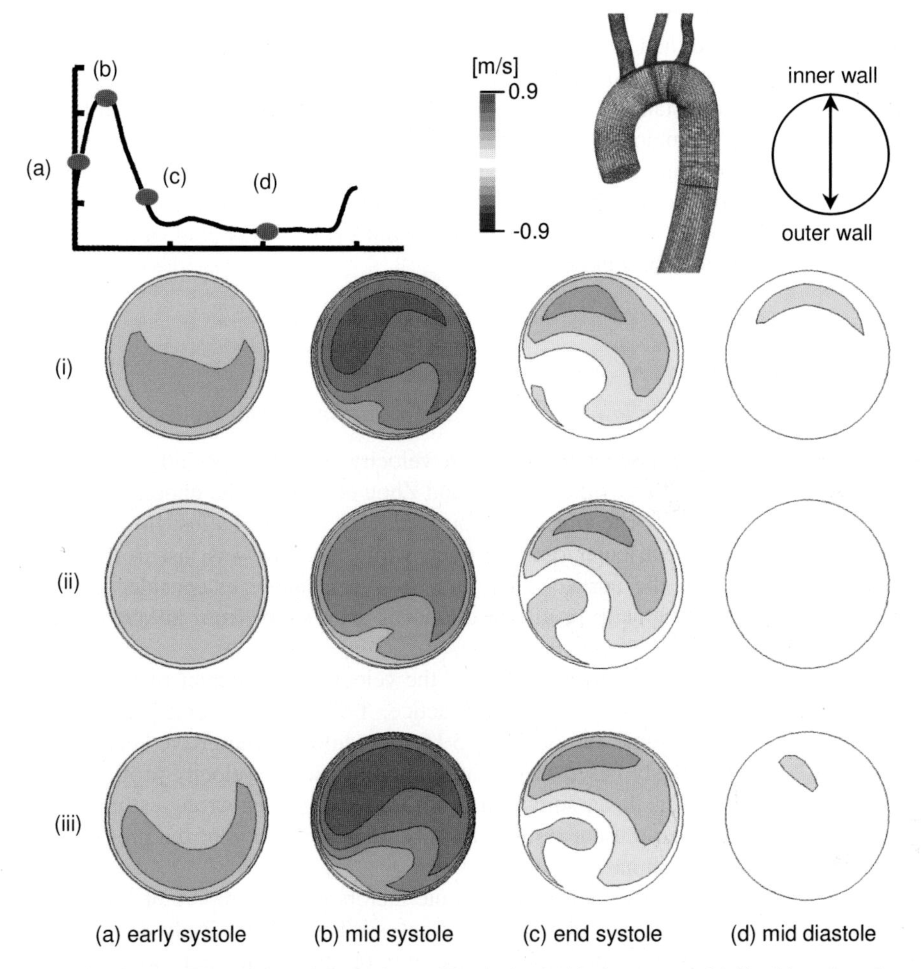

Figure 8. Contour plots of the axial velocity profile obtained with the aorta model (i) without branches and taper (upper row), (ii) with branches (middle row) and (iii) with branches and taper (bottom row). (a) early systole (left column), (b) mid-systole, (c) end systole and (d) mid diastole (right column). The time instants (a)-(d) are indicated on a time course of a flow rate curve on the left. A positive flow indicates the flow going downward. The cross-section of these contour plots and the orientation of figures are also shown.

branches, particularly posteriorly. This tendency was found throughout systole. In diastole, the WSS was almost zero and was too small for comparison.

The contour plots of the OSI are illustrated in Fig. 7. The pattern was similar overall, although differences were observed at the roots of the branches and the inner curvature of the descending aorta. The presence of branches caused a high

OSI at the roots of the branches and the left subclavian artery. The model with branches but no taper had a relatively high OSI region at the descending aorta.

Figure 8 shows contour plots of the axial velocities at the descending aorta at four different time instants at (a) early systole, (b) mid systole, (c) end systole and (d) mid diastole. Common to all models, the axial velocity was high near the outer wall in early systole and near the inner wall in mid and end systole. A regurgitant flow was also commonly observed near the outer wall in end systole. During diastole, flow velocity was almost none. The axial velocity was comparatively low in the model with branches and no taper.

4 Discussion

4.1 Significance of the inflow condition on the aortic hemodynamics

The aortic inflow did not have a uniform velocity profile, especially from mid-systole. According to Rossvoll et al. [15] and Zhou et al. [16], the nonuniformity of the blood inflow at the aortic annulus. Zhou et al. [16] observed a fast flow near the anterior wall at the ventricular outflow tract, while the flow was fast in the same sector at the inlet of the aorta. Therefore, it is reasonable to consider that the nonuniformity of the velocity profile at the aortic inlet stems from intraventricular flow.

The stoke volume is estimated by taking the velocity at the center of the inlet as the spatially averaged velocity in clinical practice. Following the same protocol, we calculated the stoke volume and obtained 59.8 ml, while it should have been 70 ml. The protocol is obviously based on the assumption of a flat velocity profile at the aortic annlus. However, this assumption is not true and gives a wrong estimation in the stroke volume. The present results strongly suggest revision of the protocol for estimating the stroke volume.

It is genratally accepted that hemodynamic factors play an important role in the development and progression of vascular diseases. The present study demostrated that the WSS distribution proximal to the top of the aortic arch was affected substantially by the aortic inflow conditions, suggesting that the occurrence of some vascular diseases at the ascending aorta may be related to a cardiac function.

4.2 Significance of the branches and tapering of the aorta

The development of a helical flow is a characteristic of aortic flow [17]. This flow phenomenon was observed in all of cases regardless of the presence of branches or a taper, indicating that the branches and taper did not qualitatively influence the helical flow.

The presence of branches gave rise to differences in the WSS and OSI at the roots of the branches. During systole, as the flow velocity was quite high in the aorta, the blood in the aorta did not flow into the branches smoothly, causing the blood to impinge on the posterior wall of the branching roots and elevating the WSS there. On the other hand, during diastole, as the WSS was almost zero during diastole, a difference between the maximum WSS and minimum WSS became large. Consequently, the OSI became higher at the roots of the branches.

Tapering of the aorta helped maintain high flow velocity in the aorta. Since this study assumed that 33% of the blood entered the branches, the flow velocity in the model with branches but no taper decreased dramatically after the branches. This was clearly reflected in the WSS distribution, and a comparison of the WSS at the distal neck of the aortic arch showed that the WSS was the lowest in this model. Therefore, although the taper did not affect the global patterns of flow in the descending aorta, it can affect the hemodynamic factors.

Published numerical studies of the human aorta have included various simplifications of the geometry. The most apparent and frequent simplification was eliminating the aortic branches, and tapering was also often ignored. With these simplifications, the aorta was reduced to the model used for the integrated model. A comparison of the results obtained in this model with those obtained in the other models showed some differences in the hemodynamic factors at the roots of the branches and the inner curvature of the descending aorta. Although the affected regions were very limited, they were sites for aneurysms and atherosclerotic lesions [18]. Therefore, from the perspective of biomedical engineering, it is important to include the branches and taper in models of the aorta to better predict the hemodynamics.

5 Conclusions

The influence of inflow conditions and geometric simplifications on the hemodynamics in the aorta was examined in a combined MRI and CFD study. The integrated flow simulation revealed that the aortic inflow did not have a flat velocity profile and brought about a difference in hemodynamics factors at the ascending aorta. A comparison of the aorta models with/without branches and tapering demonstrated that they gave rise to differences in the WSS and OSI distributions, although the affected regions were very limited and the global patterns of flow in the aorta was almost the same. These results addressed the importance of defining of inflow conditions and modeling of aortic branches for a detailed analysis of the aortic hemodynamics, charting future directions of combined MRI-CFD flow analysis.

Acknowledgments

This work was supported by the 21st Century COE Program Special Research Grant of the "Future Medical Engineering Based on Bio-nanotechnology."

References

1. Sakariassen, K.S., Barstad, R.M., 1993. Mechanisms of thromboembolism at arterial plaques, Blood Coagul. Fibrinolysis. 4, 615-625.
2. Nerem, R.M., 1993. Hemodynamics and the vascular endothelium, J. Biomech. Eng. 115, 510-514.
3. Niwa, K., Kado, T., Sakai, J., Karino, T., 2004. The effects of a shear flow on the uptake of LDL and acetylated LDL by an EC monoculture and an EC-SMC coculture, Ann. Biomed. Eng., 32, 537-543.
4. Kleinstreuer, C., Hyun, S., Buchanan, J.R. Jr., Longest, P.W., Archie, J.P. Jr., Truskey, G.A., 2001, Hemodynamic parameters and early intimal thickening in branching blood vessels, Crit. Rev. Biomed. Eng., 29, 1-64.
5. Fujioka, H., Tanishita, K., 2000. Computational fluid mechanics of the blood flow in an aortic vessel with realistic geometry, In: Yamaguchi, T. (Ed.), Clinical application of computational mechanics to the cardiovascular system, Springer-Verlag, Tokyo, pp. 99-117.
6. Mori, D., Yamaguchi, T., 2002. Computational fluid dynamics modeling and analysis of the effect of 3D distortion of the human aortic arch, Comp. Meth. Biomech. Biomed. Eng., 5, 249-260.
7. Morris, L., Delassus, P., Callanan, A., Walsh, M., Wallis, F., Grace, P., McGloughlin, T., 2005, 3-D numerical simulation of blood flow through models of the human aorta, J. Biomech. Eng., 127, 767-775.
8. Yokosawa, S., Nakamura, M., Wada, S., Isoda, H., Takeda, H., Yamaguchi, T., 2005. Quantitative measurements on the human ascending aortic flow using 2D cine phase-contrast magnetic resonance imaging, JSME Int. J. Ser. C, 48, 459-467.
9. Nakamura, M., Wada, S., Yamaguchi, T., 2006. Computational analysis of blood flow in an integrated model of the left ventricle and the aorta, AMSE J. Biomech. Eng., 128, 837-843.
10. Honma, H., Oobayashi, K., Ueda, K. 1998. Echographic anatomy for echocardiograms, In: Japan Medical Association (Ed.), Guide to Echocardiography, Nakayama, Tokyo, pp. 3-16.
11. Nakamura, M., Wada, S., Yamaguchi, T., 2006. Influence of the opening mode of the mitral valve orifice on intraventricular hemodynamics, Ann. Biomed. Eng., 34, 927-935.
12. Nakamura, M., Wada, S., Mikami, T., Kitabatake, A., Karino, T., 2003. Computational study on the evolution of an intraventricular vortical flow during early diastole for the interpretation of color M-mode Doppler echocardiograms, Biomech. Model. Mechanobiol., 2, 59-72.
13. Saber, N.R., Wood, N.B., Gosman, A.D., Merrifield, R.D., Yang, G.Z., Charrier, C.L., Gatehouse, P.D., Firmin, D.N., 2003. Progress towards patient-specific computational flow modeling of the left heart via combination of

magnetic resonance imaging with computational fluid dynamics, Ann. Biomed. Eng., 31, 42-52.
14. Buchanan, J.R., Jr., Kleinstreuer, C., 1998. Simulation of particle-hemodynamics in a partially occluded artery segment with implications to the initiation of microemboli and secondary stenosis, J. Biomech. Eng., 120, 446-454.
15. Rossvoll, O., Samstad, S., Torp, H.G., Linker, D.T., Skjaerpe, T., Angelsen, B.A., Hatle, L., 1991. The velocity distribution in the aortic anulus in normal subjects: a quantitative analysis of two-dimensional Doppler flow maps, J. Am. Soc. Echocardiogr., 4, 367-378.
16. Zhou, Y.Q., Faerestrand, S., Matre, K., Birkeland, S., 1993, Velocity distributions in the left ventricular outflow tract and the aortic annulus measured with Doppler colour flow mapping in normal subjects, Eur. Heart J., 14, 1179-1188.
17. Kilner, P.J., Yang, G.Z., Mohiaddin, R.H., Firmin, D.N., Longmore, D.B. 1993. Helical and retrograde secondary flow patterns in the aortic arch studied by three-directional magnetic resonance velocity mapping, Circulation, 88, 2235-2247.
18. DeBakey, M.E., Lawrie, G.M., Glaeser, D.H., 1985. Patterns of atherosclerosis and their surgical significance, Ann. Surg., 201, 115-131.

A FLUID-SOLID INTERACTIONS STUDY OF THE PULSE WAVE VELOCITY IN UNIFORM ARTERIES

T. FUKUI, Y. IMAI, K. TSUBOTA, T. ISHIKAWA, S. WADA AND T. YAMAGUCHI

Department of Bioengineering and Robotics, Tohoku University,
6-6-01 Aoba-yama, Sendai 980-8579, Japan
E-mail: fukui@pfsl.mech.tohoku.ac.jp

K. H. PARKER

Physiolosical Flow Studies Group, Department of Bioengineering, Imperial College, London, SW7 2AZ, United Kingdom

Pulse Wave Velocity (PWV) is recognized by clinicians as an index of mechanical properties of human blood vessels. This concept is based on the Moens-Korteweg equation, which describes the PWV in ideal elastic tubes. However, measured PWV of real human blood vessels cannot be always interpreted by the Moens-Korteweg equation because this formula is not precisely applicable to living blood vessels. It is important to understand the wave propagation in blood vessels for a more reliable diagnosis of vascular disease. In this study, we modeled uniform arteries in a three-dimensional coupled fluid-solid interaction computational scheme, and analyzed the pulse wave propagation. A commercial code (Radioss, Altair Engineering) was used to solve the fluid-solid interactions. The governing equations were the compressive Navier-Stokes equations and the equation of continuity for the fluid region, and the equation of equilibrium for the solid region. At the inlet, a steady flow with Reynolds number 1000 was imposed as the basic flow, then a single rectangular pulse with Reynolds number 4000 was imposed upon the basic flow to produce a propagating wave. We compared the PWV values obtained from the computation with those from the Moens-Korteweg equation, and showed the possibility of applying the computational technique to wave propagation analysis in human large arteries.

1 Introduction

The diagnosis of cardiovascular disease by measuring pulse wave velocity (PWV) is believed to be a promising technique. The PWV is defined as the velocity of an arterial wall disturbance toward the periphery, which occurs, for example, due to contraction of the ventricle. In general, the PWV is determined by measuring the time delay of the waveforms, Δt, between the two sites with a known distance L. Therefore,

$$PWV = \frac{L}{\Delta t}. \tag{1}$$

The PWV is believed by clinicians to be increased with the severity of vascular disease such as atherosclerosis [1-3]. The concept for applying the PWV as an index of vascular disease is based on the Moens-Korteweg equation [4], which formulates the PWV of a long straight elastic tube. According to the Moens-Korteweg equation,

$$PWV = \sqrt{\frac{Eh}{2\rho r_i}}, \tag{2}$$

where E is the Young's modulus of the arterial wall, h is the wall thickness, ρ is the blood density, and r_i is the internal radius of the artery. For thick-walled tubes, the Moens-Korteweg equation has been modified by computing the strain on the middle wall of the tube [5].

$$PWV' = \sqrt{\frac{Eh}{2\rho(r_i + h/2)}}. \tag{3}$$

In the presence of flow, we assume that the wave will be convected with the cross-sectional averaged velocity of the blood [6]. For a thick-walled tube with flow, we therefore use the "modified Moens-Korteweg equation,"

$$PWV'' = \sqrt{\frac{Eh}{2\rho(r_i + h/2)}} + U. \tag{4}$$

where U is the cross-sectional averaged velocity of the blood. It is recognized that an increased Young's modulus E will result in an increased PWV. This interpretation, however, is not always applicable to living blood vessels because the Moens-Korteweg equation includes some assumptions that are not valid for human blood vessels. The Moens-Korteweg equation is valid only when an infinitely long, straight and mechanically as well as geometrically homogeneous tube whose wall is very thin is filled with a still, non-viscous fluid. In a real artery, however, the anatomy and the constitution of the blood vessel differ from place to place, therefore, the mechanical properties of the arterial wall depend on its regional position [7]. Moreover, the geometry of the blood vessel is not infinitely long and straight but distributed complicatedly in a three-dimensional space, including many branches, curved regions, and tapering toward the periphery. In addition, the blood is not stationary but flows with its velocity changing in time and space. Hence, the diagnosis of cardiovascular disease by measuring PWV, which relies on the Moens-Korteweg equation, is not correct in the strict sense. The measured PWV of the human blood vessel is a result of several superimposed factors which influence each other, and are not so simple that the Moens-Korteweg equation can be applied. It is necessary to understand the wave propagation in blood vessels so as to make the measurements of the PWV a more reliable diagnosis of vascular disease.

In this study, we modeled uniform arteries in a three-dimensional fully coupled fluid-solid interaction computational scheme, and analyzed the pulse wave propagation. Then we estimated the PWV of long uniform arteries to assess the accuracy of our computation.

2 Methods

Pulse wave propagation can be described as an arterial wall disturbance caused by the ejection of the blood from the heart that propagates mainly toward the periphery. In this study, we imposed a steady flow as the basic flow perpendicular to the cross section of a three-dimensional long straight artery, then, a single pulse was imposed to produce a compression wave, which induced arterial wall displacement that propagated toward the periphery, and assessed the speed of the pulse wave propagation. In the uniform artery study, we compared the PWV values obtained from the computation with those from the Moens-Korteweg equation. In the stenosed artery study, a comparison was done regionally between the PWV values obtained from our computation and those from the Moens-Korteweg equation to evaluate the application of the Moens-Korteweg equation to non-ideal vessels.

2.1 Numerical models

The blood vessels that we are interested in are the relatively large and thick arteries such as the aorta. The internal radius r_i and the wall thickness h of the human aorta are approximately 10 mm and 2 mm, respectively. To meet the requirements for "sufficiently long," the axial length of the model L was set to 1000 mm, which is 100 times as long as the internal radius of the human aorta. The Young's modulus of the human aorta has been studied by many researchers [8-10]. We put a special emphasis on the Young's modulus at zero strain, and the arterial wall of our computation was assumed to be linearly elastic, with density ρ^s = 1000 kg/m^3 and Poisson's ratio ν = 0.45. The blood was assumed to be very slightly compressible to stabilize our computation. The density of the blood ρ^f is taken to be

$$\rho^f = \rho_0^f + \frac{p}{c^2}, \tag{5}$$

where ρ^f_0 is the initial density, p is the pressure relating to the external pressure, and c is the sound speed. This equation shows the relationships between the blood density ρ^f, the pressure p, and the sound speed c due to its compressibility. Higher sound speeds correspond to increased incompressibility of the fluid. The initial density ρ^f_0 was set to 1000 kg/m^3 and the sound speed c was to 60 m/s (see discussion below). The viscosity coefficient of the blood μ was set to 4.0×10^{-3} Pa·s. The different Young's moduli of the arterial wall E and the wall thickness h used in our calculations are summarized in Table 1; 9 models were used in total.

The cross section of the computational model was defined to be in the Z-X plane, and the Y-axis was in the longitudinal direction of the model. A schematic view of the uniform artery is shown in Fig. 1(a). Figure 1(b) shows one-fourth of the cross sections of the symmetrical model, and the bold line represents the border of the fluid and solid regions. The numbers of elements in one cross section were 476 for the fluid region and 224 for the solid region. The resolution along the longitudinal direction was at 5 mm intervals. The total number of elements in the model, including two extra cross sections at the inlet and the outlet, was 141,400.

The grids for the solid region, including the border of the fluid and solid regions, were solved by the Lagrangian method in order to adapt for the wall displacement due to the disturbance by the pulse. The grids for the fluid region except for the cells on the fluid-solid border were fixed and solved by the Eulerian method.

Table 1. Young's modulus and wall thickness of the model uniform artery.

Young's modulus E (MPa)	0.5	1.0	2.0
Wall thickness h (mm)	1.5	2.0	4.0

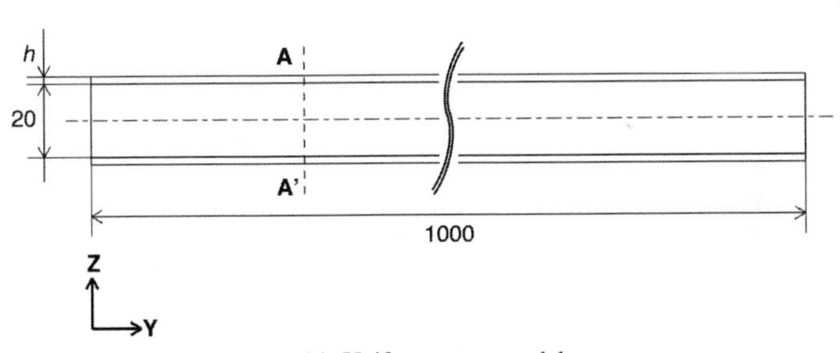

(a) Uniform artery model.

(b) Cross section.

Figure 1. Schematic views of the artery models used in this study. (a) shows a uniform artery model, (b) shows one-fourth of the cross sections of the symmetrical model. The bold line in the cross section represents the border of the fluid and solid regions. The number of elements in one cross section; 476 for the fluid, 224 for the solid.

2.2 Governing equations and computational code

A commercial code (Radioss, Altair Engineering) was used to solve the fluid-solid interactions with an ALE formulation. The ALE stands for Arbitrary Lagrangian Eulerian, which enables us to solve the interactions with Lagrangian and Eulerian methods arbitrarily. The governing equations for the compressible fluid were the Navier-Stokes equations (Eq. (6)) and the equation of continuity (Eq. (7)).

$$\frac{\partial(\rho^f u_j^f)}{\partial t}+\frac{\partial(\rho^f u_i^f u_j^f)}{\partial x_i}=-\frac{\partial p}{\partial x_j}+\mu\frac{\partial^2 u_j^f}{\partial x_i \partial x_i}, \tag{6}$$

$$\frac{\partial \rho^f}{\partial t}+\frac{\partial(\rho^f u_i^f)}{\partial x_i}=0, \tag{7}$$

where f superscribed on the parameters indicates the fluid region, ρ^f is the fluid density, u_i^f is the velocity, p is the pressure, and μ is the viscosity coefficient. The equation of equilibrium (Eq. (8)) was solved for the solid wall. The solid was assumed to be a linearly elastic solid (Eq. (9)).

$$\int_V \frac{\partial \sigma_{ij}}{\partial x_j} \mathrm{d}V = \int_V \rho^s \frac{\partial u_i^s}{\partial t} \mathrm{d}V, \tag{8}$$

$$\sigma_{ij} = C_{ijkl}\varepsilon_{kl}, \tag{9}$$

where s superscribed on the parameters indicates the solid region, σ_{ij} is the Cauchy stress tensor, C_{ijkl} is the elastic tensor, and ε_{kl} is the strain tensor.

2.3 Boundary conditions

The boundary conditions were that both ends of the artery were fixed (Eq. (10)), no-slip on the wall (Eq. (11)), and the 'silent boundary' condition which enables us to reduce reflections at the outlet (Eq. (12)).

$$u_i^s = \omega_i^s = 0, \tag{10}$$

$$u_i^f = u_i^s, \tag{11}$$

$$\frac{\partial p}{\partial t} = \rho^f c \frac{\partial u_n^f}{\partial t} + c\frac{(p_\infty - p)}{2l_c}, \tag{12}$$

where ω is the angular velocity, n is the vector perpendicular to the cross section, p_∞ is the reference pressure, and l_c is a length characteristic of the grid size. At the inlet, a steady uniform velocity, with Reynolds number 1000 (= 0.2 m/s), perpendicular to the cross section was imposed as the basic flow and computations continued until the initial oscillations of the wall reduced to less than 0.1% of the internal radius of the artery. Then, a single rectangular pulse with a period of 10 ms and the Reynolds number of 4000 (= 0.8 m/s) was imposed upon the basic flow to produce a propagating wave (Fig. 2). The time to establish the basic flow depended on the artery models, however, it was around 10 (s) which is 1000 times as long as that of the single pulse at t = 0.0 (ms). Then, the velocity, displacement, and pressure values were obtained with a sampling rate of 1000 Hz in each model to draw waveforms.

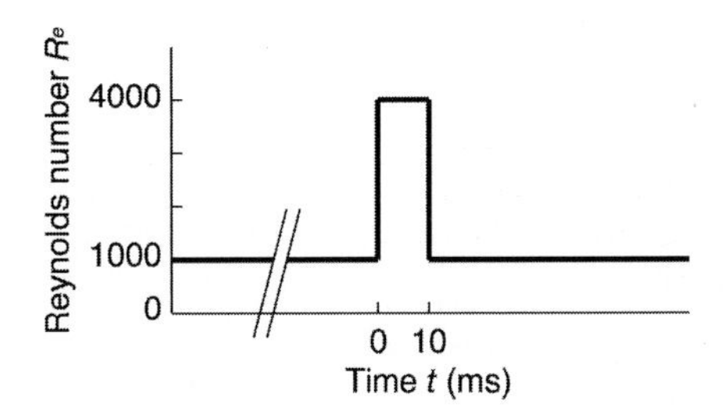

Figure 2. Boundary condition at the inlet. A steady flow was imposed before a single pulse as the basic flow. The time taken to establish the basic flow was around 10 s.

3 Results

For the models used in this study, it took about three weeks to finish each calculation with a Linux machine whose processor and memory were 2.4 GHz and 1 GB.

3.1 Wave propagation

Figure 3 shows the color-coded distribution of the axial velocity u_y in the uniform artery at different times with the Young's modulus of 1.0 MPa and the wall thickness of 2.0 mm at the plane x = 0. The scale in the radial direction is multiplied by 4 for clarity. The propagation of the velocity pulse towards the periphery is clear.

3.2 Velocity waveforms

The velocity waveforms in the uniform artery were obtained from the axial velocity u_y on the center-line of the artery. Figure 4 shows the velocity waveforms in the uniform artery with the Young's modulus of 1.0 MPa and the wall thickness of 2.0 mm at y = 350, 500, and 650 mm. The peak of the waveform shifted peripherally in time, indicating this wave was forward-going. The PWV was calculated from the estimate of the time delay Δt of each waveform and the distance between the two sites of measurement. This is generally known as the foot-to-foot method [11] since the reference point of the waveform is the foot of each waveform. The foot point was determined by the intersection of the two straight lines; the non-oscillating line before the waveform and the tangential line of the ascending slope. The tangential line was determined to approximate the data between 20 ~ 80% of the ascending slope by least squares. The PWV was then calculated regionally at intervals of 100 mm from y = 50 ~ 950

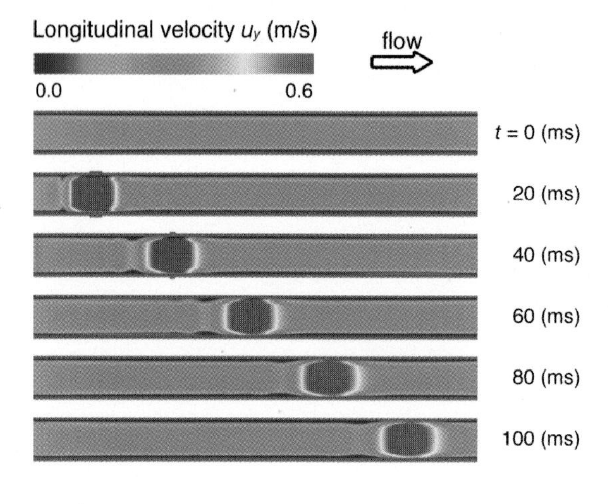

Figure 3. Color-coded distribution of the longitudinal velocity u_y in the uniform artery with the Young's modulus of 1.0 MPa and the wall thickness of 2.0 mm at the plane x = 0. The scale in the radial direction is multiplied by 4 for clarity.

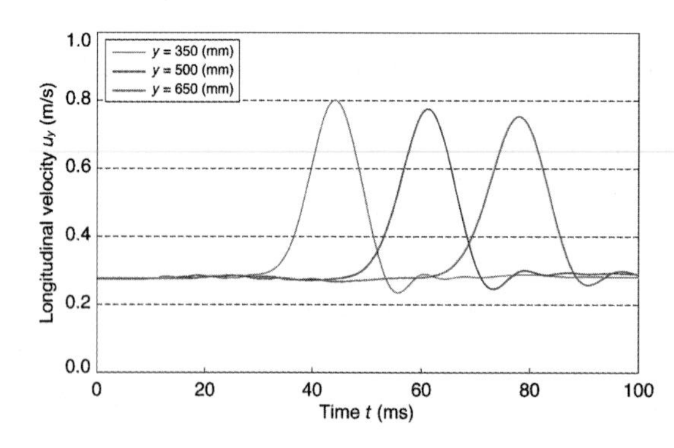

Figure 4. Center-line velocity waveforms at different locations in the uniform artery with the Young's modulus of 1.0 MPa and the wall thickness of 2.0 mm.

3.3 Pulse wave velocity

Figure 5 presents the regional PWV of uniform arteries calculated with a Young's modulus of 1.0 MPa. The PWV at y = 100 mm was obtained from the waveforms at y = 50 and 150 mm. The PWV at y = 200 mm was obtained from the waveforms at y = 150 and 250 mm, and so on. The differences in color show the differences in the wall thickness of the artery. The dotted lines show the PWV estimation by the modified Moens-Korteweg equation (Eq. (4)). The averaged velocity over a cross section U in Eq. (4) was determined with the flow velocity values at $t = 0.0$ ms. There were good correspondences between the values from the computation and those from the modified Moens-Korteweg equation for the arteries with wall thicknesses of 1.5 and 2 mm. For the artery with a wall thickness of 4 mm, the values were significantly lower than that from the modified Moens-Korteweg equation. There was variation in the PWV values in the longitudinal direction, though, it was less than 10% of the PWV. The larger variation can be seen near both ends. The differences decreased to less than 5% when the PWV values between $y = 300$ to 700 mm were used.

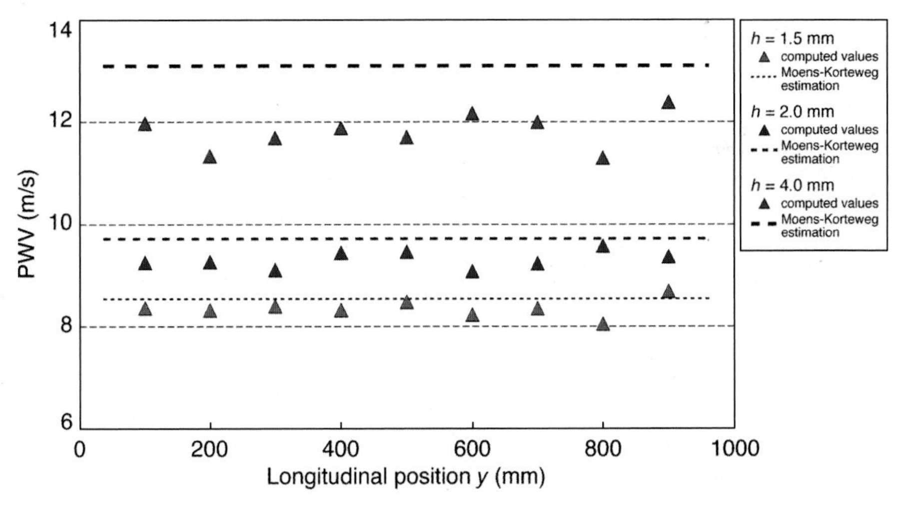

Figure 5. PWV values of the uniform arteries with the Young's modulus of 1.0 MPa in the longitudinal direction. The dotted lines show the PWV estimation by the modified Moens-Korteweg equation.

3.4 PWV comparison between computation and theoretical values

The PWV of the nine uniform arteries were estimated, and compared with those from the modified Moens-Korteweg equation (Fig. 6). The ordinate shows the PWV obtained from the computation at the center of the artery, $y = 500$ mm, and the abscissa shows those from the modified Moens-Korteweg equation (Eq. (4)). The color denotes the wall thickness of the artery and the shape denotes the Young's modulus of the arterial wall. The dotted line indicates the line of equality between the two parameters. The PWV from the computation was lower than those from the modified Moens-Korteweg equation in the higher range of the PWV.

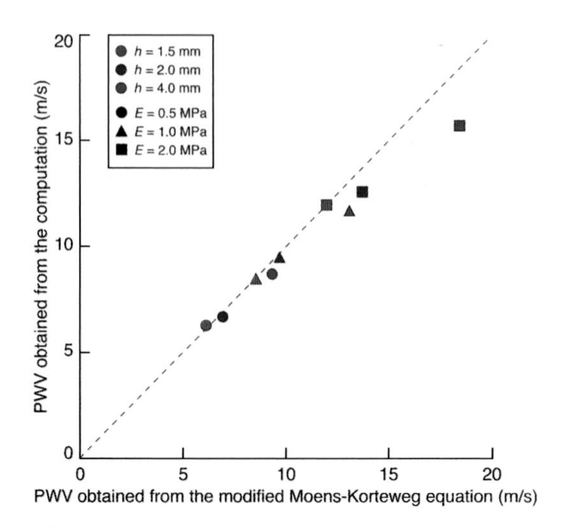

Figure 6. Comparison of the PWV obtained from the computation and those obtained from the modified Moens-Korteweg equation. The dotted line indicates the line of equality between the two parameters.

4 Discussion

As shown in Fig. 5, the PWV values were not constant in the longitudinal direction, but varied by up to 10%. The variation can be seen mainly at both ends and could be attributed to the boundary conditions of our computation. The boundary conditions at both ends were fixed, which means no translating or rotating motions at the ends. Due to the fixed translating motion, the arterial wall at both ends could not be dilated in the radial direction nor elongated in the longitudinal direction. These fixed conditions would produce reflection which could affect the PWV values at the ends. We concluded that when a uniform artery with a length of 1000 mm is used in our computation, the PWV estimation is valid in the center 500 mm of the artery, where the variation of the PWV even for the thickest-walled vessel was less than 5%.

In the higher range of the PWV, the PWV obtained from the computation were lower than those from the modified Moens-Korteweg equation as shown in Fig. 6. One of the reasons for this difference is that the Moens-Korteweg equation is only valid for a cylinder with a very thin wall. The differences between the two parameters were up to 3.9% for the wall thickness of 1.5 mm, up to 8.0% for 2.0 mm, and up to 16.1% for 4.0 mm. The most significant differences can be seen for the wall thickness of 4.0 mm (i.e., 20% of the diameter).

Another reason for the differences in the computation is the inclusion of the sound speed of the blood. The Young's modulus of the arterial wall and the bulk modulus of the blood differ by 10^3 to 10^4 if the sound speed of the blood is assumed to be the more realistic value of 1500 m/s and the bulk modulus K can be calculated as $K = \rho c^2$. It is impracticable to deal with materials whose elastic modulus differ by 10^3 to 10^4 with each other for the fluid-solid interaction study with current computer techniques. In this study, therefore, the elastic modulus of the fluid was decreased to the order of that of the solid by assuming the fluid was slightly compressible with the sound speed of 60 m/s. The underestimated PWV of the uniform artery for the higher range of the PWV was attributed to the decreased bulk modulus of the blood. The PWV, which can be expressed as the wave speed due to radial displacement of the arterial wall, never exceeds the sound wave speed. The higher PWV was more affected by the compressibility of the blood and more underestimated. Nevertheless, the differences between the PWV obtained from the computation and those from the modified Moens-Korteweg equation were less than 7% up to 12 m/s of the PWV, indicating these computational methods for the PWV analysis were accurate enough to evaluate its value quantitatively. Moreover, this range is similar to the PWV of the human aorta, which is our greatest interest. These results show the possibility of applying the computational technique to wave propagation analysis in human large arteries.

5 Conclusions

In conclusions, we analyzed the pulse wave propagation with a three-dimensional fluid-solid interaction computational scheme, and showed the possibility of applying the computational technique to wave propagation analysis in human large arteries.

Acknowledgments

This research was supported by the following grants;
"Revolutionary Simulation Software (RSS21)" project supported by next-generation IT program of Ministry of Education, Culture, Sports, Science and Technology (MEXT).
Grants in Aid for Scientific Research by the MEXT and JSPS
Scientific Research in Priority Areas (768) "Biomechanics at Micro- and Nanoscale Levels"
Scientific Research(A) No.16200031 "Mechanism of the formation, destruction, and movement of thrombi responsible for ischemia of vital organs"

References

1. Ting, C.T., Chou, C.Y., Chang, M.S., Wang, S.P., Chiang, B.N., Yin, F.C., 1991. Arterial hemodynamics in human hypertension. Effects of adrenergic blockade, Circulation 84, 1049-1057.
2. Lehmann, E.D., Gosling, R.G., Sonksen, P.H., 1992. Arterial wall compliance in diabetes, Diabetic Med. 9, 114-119.
3. Triposkiadis, F. Kallikazaros, I. Trikas, A. Stefanadis, C. Stratos, C. Tsekoura, D. Toutouzas, P., 1993. A comparative study of the effect of coronary artery disease on ascending and abdominal aorta distensibility and pulse wave velocity. Acta Cardiologica 48, 221-233.
4. Bramwell, J.C., Hill, AV., 1922. The velocity of the pulse wave in man, Proceedings of the Royal Society, London B93, pp. 298-306.
5. Fung, Y.C., 1996. Biomechanics Circulation (2nd Ed.), Springer, New York, pp. 140-144.
6. Khir, A.W., O'Brien, A., Gibbs, J.S.R., Paker, K.H., 2001. Determination of wave speed and wave separation in the arteries, J. Biomech. 34, 1145-1155.
7. Nakashima, T., Tanikawa, J., 1971. A study of human aortic distensibility with relation to atherosclerosis and aging, Angiology 22, 477-490.
8. MacSweeney, S.T., Young, G., Greenhalgh, R.M., Powell, J.T., 1992. Mechanical properties of the aneurysmal aorta, Br. J. Surg. 79, 1281-1284.
9. He, C.M., Roach, M.R., 1994. The composition and mechanical properties of abdominal aortic aneurysms, J. Vasc. Surg. 20, 6-13.
10. Thubrikar, M.J., Labrosse, M., Robicsek, F., Al-Soudi, J., Fowler, B., 2001. Mechanical properties of abdominal aortic aneurysm wall, J. Med. Eng. Technol. 25, 133-142.
11. Khir, A.W., Zambanini, A., Parker, K.H., 2004. Local and regional wave speed in the aorta: effects of arterial occlusion, Med. Eng. Phys. 26, 23-29.

RULE-BASED SIMULATION OF ARTERIAL WALL THICKENING INDUCED BY LOW WALL SHEAR STRESS

S. WADA AND M. NAKAMURA

Department of Mechanical Science and Bioengineering,
Graduate School of Engineering Science, Osaka University,
1-3 Machikaneyama, Toyonaka, Osaka 560-8531, Japan
E-mail: shigeo@mech.es.osaka-u.ac.jp

T. KARINO

Research Institute for Electronic Science,Hokkaido University,
North12, West 6, North District, Sapporo 060-0812, Japan

To investigate the effects of hemodynamic factors on the development of atherosclerosis and intimal hyperplasia, we carried out a computer simulation of the adaptive changes in the thickness of the wall of a human coronary artery with multiple bends. This was done by assuming that only the arterial wall where wall shear stress (WSS) is lower than a certain threshold value increases its thickness, and shifting the luminal surface of the vessel step by step in the direction normal to the wall based on the value of WSS obtained by the calculation of blood flow through the artery under conditions of a steady flow. It was found that thickening of the vessel wall occurred and progressed at the inner wall of curved segments where WSS was low at the initial state. However, the final thickness of the wall at the completion of adaptive changes was not determined by the value of WSS in the initial state, but was determined as a result of the interaction between a change in vascular geometry caused by thickening of the vessel wall and the flow affected by the vascular geometry.

1 Introduction

Atherosclerotic lesion and intimal thickening are observed more commonly at a bifurcation and a curved segment of a large or middle sized artery in the human vascular system [1-2]. Fluid dynamics researches have revealed that the blood flow is disturbed and wall shear stress (WSS) is relatively low at the favorite sites of atherosclerosis [3-6]. Therefore it is suspected that hemodynamic factors are involved in the development of atherosclerosis, in particular, flow-induced WSS plays an important role in the localized pathogenesis. However, most of the flow studies investigate the arterial blood flow after the atherosclerotic lesion have already developed in order to relate the WSS to atherogenesis. Thickening of the arterial wall by the development of atherosclerosis largely changes the geometry of the blood vessel. It would affect the blood flow in the artery, resulting in the hemodynamic factors involved in the development of atherosclerosis. In order to find out the role of hemodynamics in the development of atherosclerosis and the mechanism of the localization of atherosclerosis, it is necessary to follow up its progression.

Computer aided analysis allows us to predict the progression of wall thickening leading to atherosclerosis based on various hypotheses. However, few studies [7-9] have been carried out to account for the change in geometry of blood vessel with the progression of wall thickening. In this study, we conducted a computer simulation of an adaptive change induced by WSS in the thickness of the wall of a human coronary artery with multiple bends.

2 Methods

2.1 Initial geometry of the blood vessel

The outline of a blood vessel with a multiple bend was obtained from a transparent human right coronary artery [6], neglecting the branches stemming from it. It was assumed that the thickness of the vessel wall was uniform everywhere in the initial state, and then the boundary of the inside wall was determined from that of the outside wall. The diameter of the vessel was 4.33 mm at the entrance of the arterial segment, and the length along the central axis was 28 mm. The artery model was constructed by assuming that the vessel was symmetrical with respect to its common median plane of multiple bend, and the cross-sections were circular at any location (Fig. 1). A hexahedral element with 8 nodes was used for computational fluid dynamics (CFD) analysis. The vessel was divided into 150 and 80 equal-sized elements in the longitudinal and circumferential directions, respectively, and 41 elements of gradually diminishing size in the radial direction.

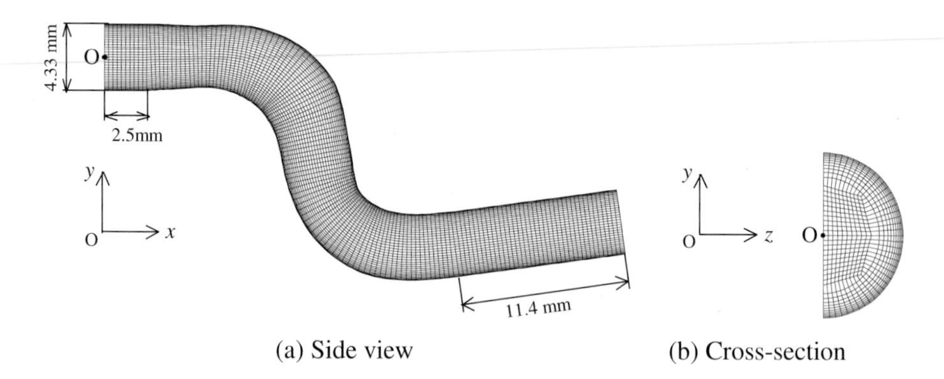

Figure 1. CFD model of a human coronary artery with multiple bend.

2.2 *Blood flow analysis*

Under the assumptions that arterial wall is rigid and blood is an incompressible Newtonian fluid with a density of 1.05×10^3 kg/m^3 and a viscosity of 3.5×10^{-3} Pa·s, the continuity and Navier-Stokes equations for steady flow were solved by the use of a flow simulation software (Star LT distributed by CD-adapco JAPAN Co., LTD). Boundary conditions applied were a parabolic velocity profile at the inlet, a uniform pressure at the outlet, and non-slip condition at the vessel wall.

2.3 *Thickening of the vessel wall*

Based on the fact that atherosclerosis likely occurs at a low wall shear stress region in the artery [3, 4], it was assumed that the vessel wall thickens inward at the location where WSS is less than a threshold value, τ_{th}. In this study, the amount of thickening during a certain period corresponding to a computational step is given by

$$\delta = \begin{cases} C\left(\tau_{th} - \tau\right)/\tau_{th} & \text{for } \tau \le \tau_{th} \\ 0 & \text{for } \tau > \tau_{th} \end{cases} \tag{1}$$

where τ is the WSS calculated from the flow velocity in the artery, and C is a coefficient (Fig. 2).

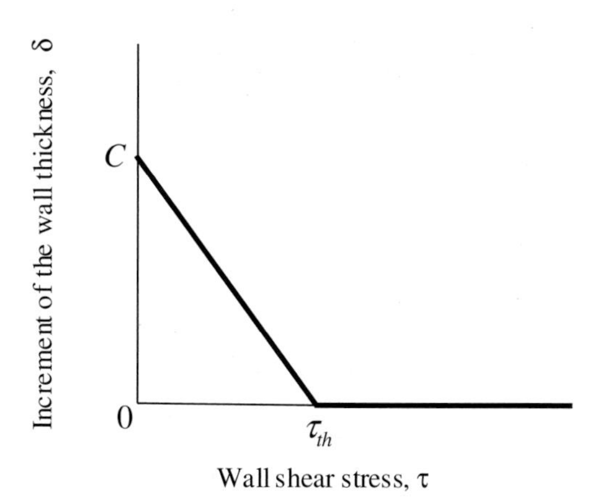

Figure 2. Relationship between wall shear stress and increment of the thickness of an arterial wall during a certain period.

2.4 Procedure of computer simulation

Computer simulations were conducted by repeating calculation of blood flow, evaluation of wall shear stress, and change in the geometry of the vessel by thickening of the wall. A new geometry of the vessel at each step was obtained by moving the nodal points on the wall by δ in Eq. (1) in the direction normal to the lumen. The solution domain was re-meshed for the new geometry. This process was repeated until a stable geometry of the vessel was obtained. The values of τ_{th} and C chosen were 1.2 Pa [10] and 100 μm, respectively.

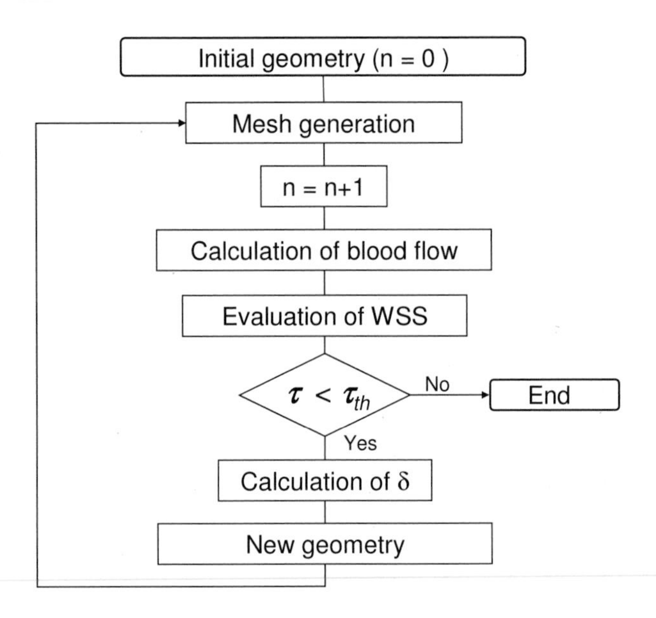

Figure 3. Procedure of the rule-based simulation of arterial wall thickening.

3 Results and Discussion

Calculations were carried out for the blood flowing at a physiological flow rate of 2.83×10^3 mm^3/s in the human coronary artery. The Reynolds number at the inlet was 250. A stable geometry of the artery was obtained in 35 computational steps.

Figure 4 shows the velocity profile of blood flow in the initial geometry of the artery before thickening of the wall occurs. The results show the axial velocity profile in the symmetric plane, and contour plots of the axial velocity and vector plots of the secondary flow in selected cross-sections. The secondary flow is viewed from the downstream. The values shown outside and inside the vessel indicate the magnitude of wall shear stress and the maximum velocity of axial flow at that location. At the first curved segment, the blood flowing around the central axis moves toward the outer wall (upper wall in this figure) of the bend by the inertial force. This produces a secondary flow rotating in a single direction in the half cross-section shown in this figure. At the second bend, the direction of the secondary flow is switched and the blood flow is much more disturbed. The complex flow patterns with a strong secondary flow creates the region of low and high wall shear stress in the multiple bend artery. The WSS is less than the threshold value at three sites; upper wall proximal to the first bend, lower wall between the first and second bends, and upper wall distal to the second bend.

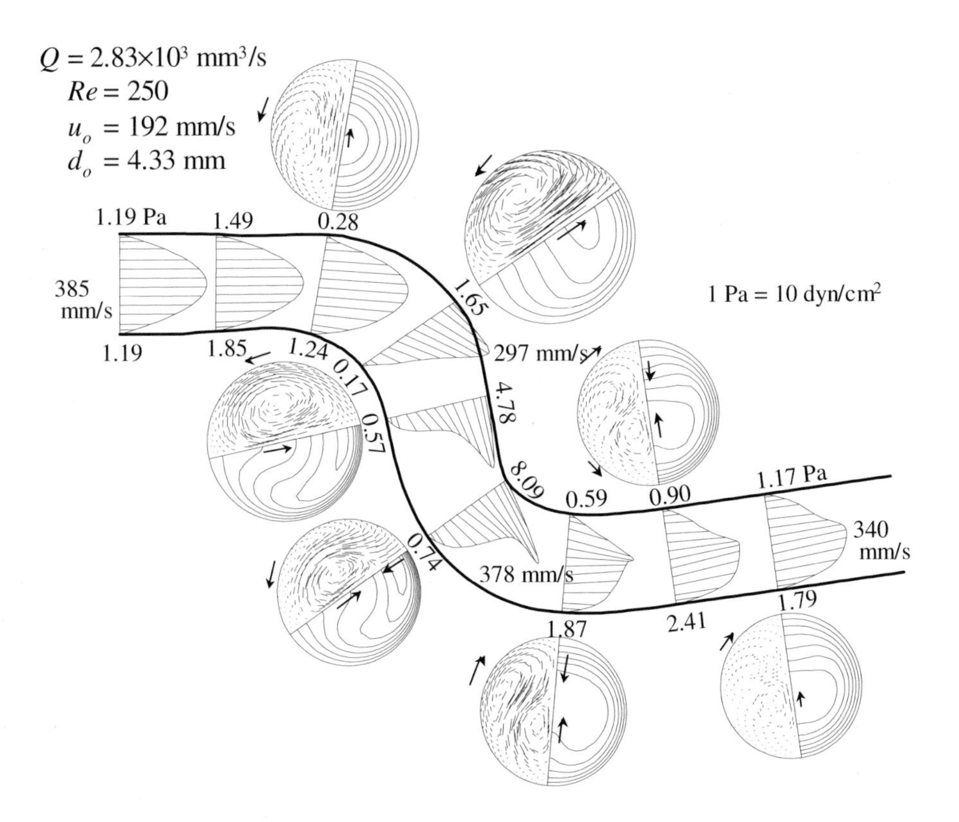

Figure 4. Distributions of fluid velocity in the plane of symmetry at a flow rate of 2.83×10^3 mm^3/s and the cross-sections of the artery in the initial geometry before thickening of the wall occurs.

Figure 5 shows the velocity profile of blood flow in the geometry of the artery obtained at the 35th step. The configuration of the cross-section is changed by thickening of the wall, and the WSS is almost greater than the threshold value everywhere. However, the axial velocity profile and secondary flow patterns do not vary significantly from the initial state before thickening of the wall. This indicates that thickening of the wall does not affect the global flow patterns in the artery but the local flow around the site where the thickening occurs.

Figure 6 shows the change in the geometry of the artery by the progression of thickening of the wall from the initial to the final stable state. The results show the lumen of the blood vessel in the symmetric plane and the cross-sections. The values shown inside and outside cross-section are the incremental thickness of the

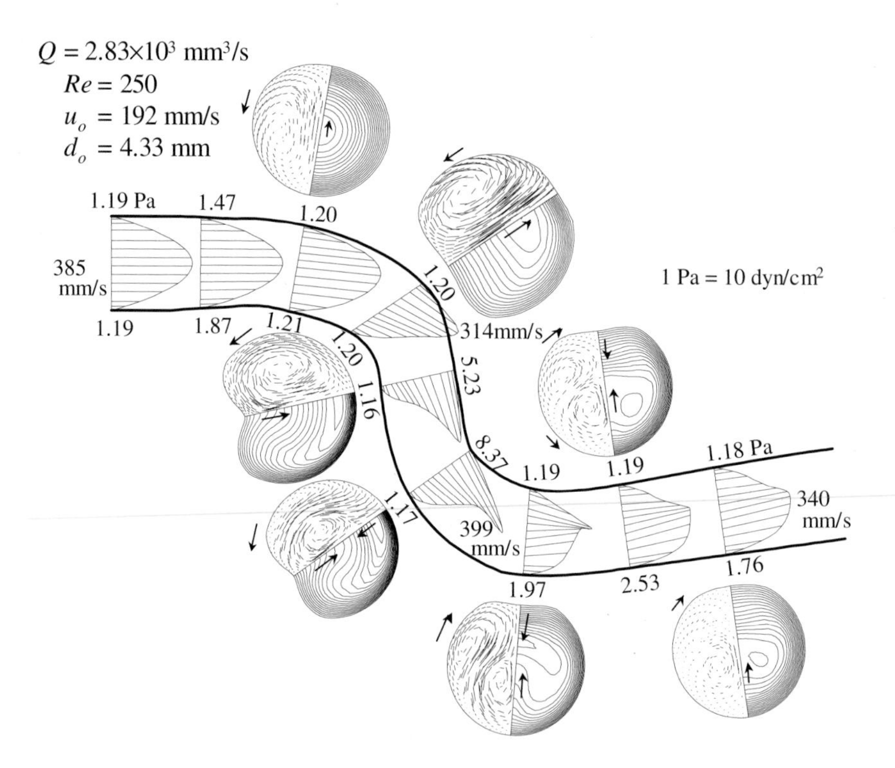

Figure 5. Distributions of fluid velocity in the plane of symmetry at a flow rate of 2.83×10^3 mm^3/s and the cross-sections of the artery after in the final geometry after thickening of the wall occurred.

wall and the WSS in the initial geometry at that location, respectively. The thickening of the wall onset from the site where the WSS is less than the threshold in the initial step and the site is extended gradually with the progression of the thickening. The maximum thickening of the wall is observed at the inner wall of the first bend where the wall shear stress is the minimum in the initial geometry. However, the vessel wall in the cross-section C is thickened more than the cross-section B although the WSS in the initial geometry is larger.

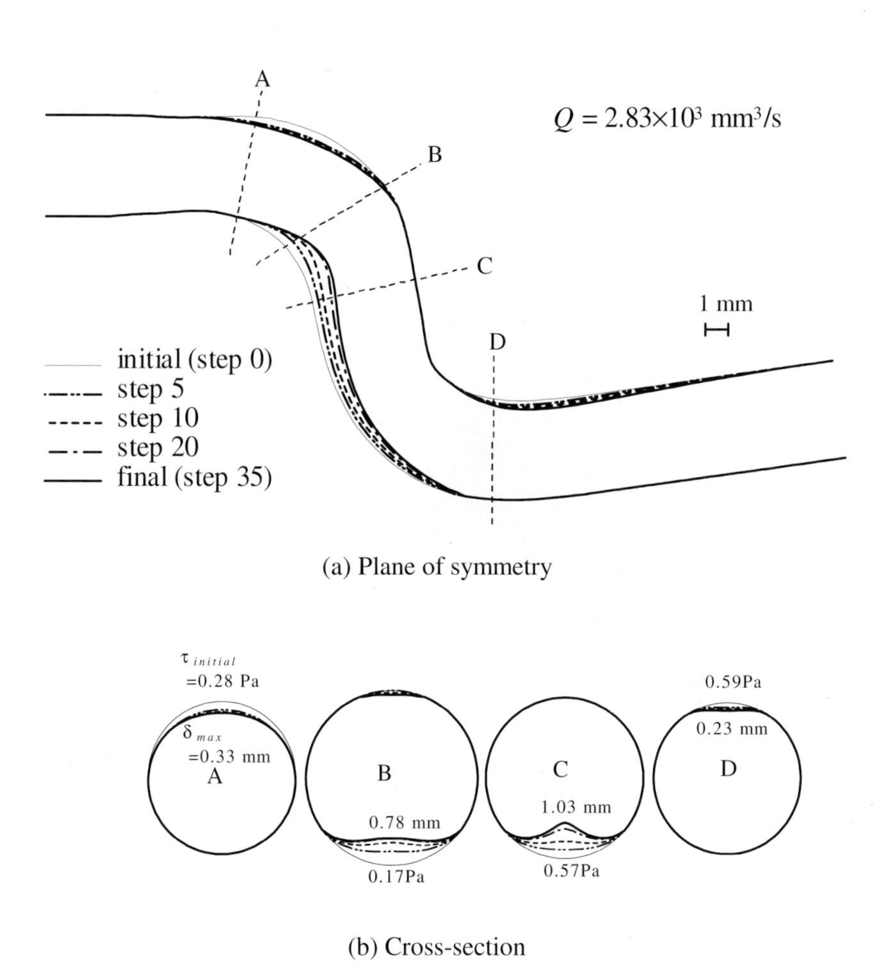

Figure 6. Change in the geometry of the arterial model with a multiple bend by the progression of wall thickening ($Q = 2.83\times10^3$ mm^3/s).

Figure 7 shows the relationship between the wall shear stress in the initial state and the increment of wall thickness in the stable state at all of the nodal points over the luminal surface of the artery. It is found that thickening occurs even at the sites where wall shear stress is larger than the threshold value in the initial geometry because change in the geometry of the blood vessel generates a new site of low wall shear stress. The increment of the wall thickness tends to increase with decreasing the wall shear stress. However, lower wall shear stress in the initial state does not always induce thicker wall. This indicates that the final thickness of the wall is determined not only by the value of wall shear stress before thickening occurs but also by the change in the flow patterns caused by the change in geometry of the blood vessel with the progression of thickening of the wall.

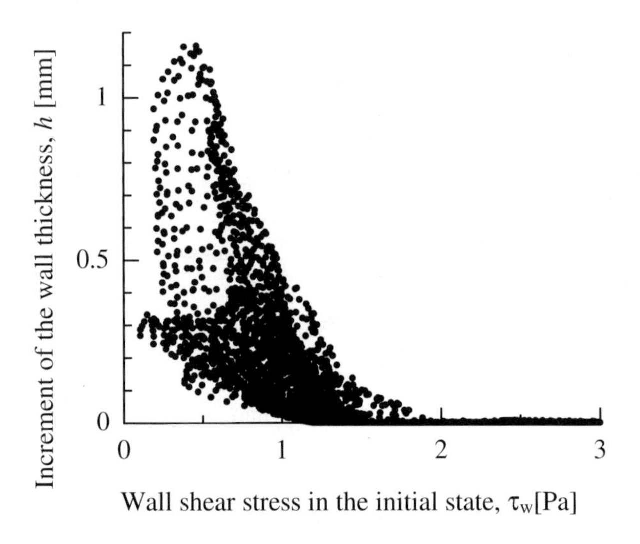

Figure 7. Relationship between the wall shear stress in the initial state and the increment of the wall thickness in the final sate.

4 Conclusions

In this paper, the progression of atherosclerosis in the human carotid artery with multiple bends was investigated by computer simulation. Based on the hypothesis that the arterial wall locally thickens at the site of low wall shear stress, the development of the wall thickening was computed including changes in the geometry of the artery and blood flow by the progression of the wall thickening. The rule-based simulation showed that the arterial wall is thickened at the site of low wall shear stress in the initial geometry before thickening occurs, but the degree of the wall thickening is not determined by only the magnitude of the wall shear stress. These results provided us a new insight into the localization mechanism of vascular diseases.

Acknowledgments

This work was supported by Grant-in-Aid for Scientific Research No. 17300138 from the Ministry of Education, Culture, Sports, Science and Technology of Japan.

References

1. Fox, B., James, K., Morgan, B., Seed, A., 1982. Distribution of fatty and fibrous plaques in young human coronary arteries, Atherosclerosis 41, 337-347.
2. Grøttum, P., Svindland, A., Walløe, L., 1983. Localization of atherosclerotic lesions in the bifurcation of the main left coronary artery, Atherosclerosis 47, 55-62.
3. Caro, C. G., Fitz-Gerald, J. M., Schroter, R. C., 1971. Atheroma and arterial wall shear. Observation, correlation and proposal of a shear dependent mass transfer mechanism for atherogenesis, Proc. Roy. Soc. Lond. B177, 109-159.
4. Zarins, C. K., Giddens, D. P., Bharadvaj, B. K., Sottiurai, V. S., Mabon R.F., Glagov, S, 1983. Carotid bifurcation atherosclerosis. Quantitative correlation of plaque localization with flow velocity profiles and wall shear stress, Circ. Res. 53, 502-514.
5. Ku, D. N., Giddens, D. P., Zarins, C. K., Glagov, S, 1985. Pulsatile flow and atherosclerosis in the human carotid bifurcation. Positive correlation between plaque location and low and oscillating shear stress, Atherosclerosis 5, 293-302.
6. Asakura, T., Karino, T., 1990. Flow patterns and spatial distribution of atherosclerotic lesions in human coronary arteries, Circ. Res. 66, 1045-1066.
7. Lee, D., Chiu, J. J., 1992. A numerical simulation of intimal thickening under shear in arteries, Biorheology 29, 337-351.
8. Lee, D., Chiu, J. J., 1996. Intimal thickening under shear in a carotid bifurcation--a numerical study, J. Biomech. 29, 1-11.
9. Ishikawa, T., Oshima, S., Yamane, R., 1999. Modeling and Numerical Simulation of Axisymmetric Stenosis Growth in an Artery (in Japanese), Trans. JSME B 65-637, 2982-2989.
10. Kamiya, A., Togawa, T., 1980. Adaptive regulation of wall shear stress to flow change in the canine carotid artery, Am. J. Physiol. 239, H14-H21.

SUBJECT INDEX

A

Actin, 43, 67, 84
Agarose, 107
Anulus fibrosus, 37
Aorta, 133, 148
Aortic annulus, 139
Aortic valve, 133
Arbitrary Lagrangian Eulerian (ALE) formulation, 150
Artery, 135, 147, 157
Articular cartilage, 107, 118
Ascending aorta, 135
Atherosclerosis, 134, 146, 157
Atomic force microscope (AFM), 15
Axial velocity, 141, 151, 160

B

Bell model, 67
Biodegradable polymers, 127
Biomembrane force probe, 76
Biomolecules, 66
Biphasic scaffold, 122
Blood flow, 81, 134, 157
Blood vessel, 81, 144, 157
Bone adaptation, 26
Branches, 17, 133, 147, 158
Brownian motion, 51, 78

C

Calcium, 13, 87, 125
Calcium signaling, 28
Cardiovascular, 146
Cartilage, 43, 81, 107, 118
Catch bonds, 51, 66
Cell adhesion, 15, 51, 66, 86, 118
Cell alignment, 40
Cell rolling, 58
Cellular network, 13
Chondrocyte, 43, 87, 107, 122
Cine phase-contrast, 133
Collagen, 37, 92, 107, 119

Compressible fluid, 150
Computational fluid dynamics (CFD), 133, 158
Confocal microscopy, 37
Conformational changes, 68
Coronary artery, 157
Cytoplasmic strain, 43
Cytoskeleton, 36, 85

D
Descending aorta, 133
Dimensional analysis, 51
DMMB assay, 112

E
E-cadherins, 87
Elastic substrates, 36
Elastic tubes, 146
Energy landscape, 77
Extracellular matrix, 14, 43, 51, 85, 123

F
Fixed charge density (FCD), 107
Flow chamber, 53, 66
Flow-enhanced cell tethering, 53
Fluid-solid interaction, 146
Focal adhesion kinase (FAK), 85
Focal adhesions, 36
Focal complex, 85
Fyn/Shc, 85

G
Gentamicin, 3
Gerbil, 3
G-protein coupled receptors (GPCR), 88

H
Heart, 81, 148
Hemodynamics, 133, 157
Hexahedral element, 158
Hybrid mesh, 119
Hybrid scaffold, 118
Hybrid sponge, 119

I

Incompressible Newtonian fluid, 136, 159
Inflow condition, 133
Integrins, 36, 85
Interdomain angle, 72
Intervertebral disc, 36
Intracellular strain, 39

J

Jun N-terminal kinase (JNK), 91

K

Kinetics, 53, 66

L

Large arteries, 146
Left ventricle, 133
Linearly elastic, 148
Localization of atherosclerosis, 158
L-selectin, 51, 67

M

Magnetic resonance imaging (MRI), 107, 133
Magnetic resonance spectroscopy (MRS), 107
Mechanical deformation, 36
Mechanical properties, 43, 123, 146
Mechanoresponse, 81
Mechanosensitive (MS) ion channels, 83
Mechanosensors, 83
Mechanotransduction, 3, 13, 36, 80
Membrane fluidity, 88
Mesenchymal stem cells (MSCs), 125
Mitochondria, 37
Mitogen-activated protein kinase (MAPK), 91
Mitral valve, 136
Moens-Korteweg equation, 146
Molecular diffusion, 56
Molecular dynamics simulations, 66
Monte Carlo, 66
Motility, 3, 30
Multiple bends, 157
Mutation, 3, 64

N
Nanoindentation, 18
Nondestructive, 107
Noninvasive, 107
Nuclear magnetic resonance (NMR), 108
Nuclear strain, 36
Nucleus, 4, 17, 36, 87

O
Off-rate, 53, 67
On-rate, 53, 75
Organelle, 36
Oscillatory shear index, 137
Osteochondral tissue, 118
Osteocyte, 13
Ototoxic, 3
Outer hair cell, 3

P
Paxillin, 85
Poly (lactic-co-glycolic acid) (PLGA), 119
Porous, 107, 118
Prestin, 3
Proteoglycan, 107
P-selectin glycoprotein ligand 1 (PSGL-1), 52, 67
Pulse wave velocity (PWV), 146

R
Regenerative medicine, 108
Rolling regularity, 61
Rule-based simulation, 157

S
Scaffold, 114, 118
Scaling, 55
Secondary flow, 133, 161
Selectin, 51, 66
sGAG, 111
Sialyl Lewis X (sLe^x), 67
6-sulfo-sLe^x, 76
Sliding velocity, 53
Sliding-rebinding mechanism, 66
Slip bonds, 66

Sound speed, 148
Steady flow, 146, 157
Strain, 36, 81, 147
Strain transfer, 36
Strain transfer ratio, 36
Streptomycin, 3
Sulfated glycosaminoglycan, 111

T
Taper, 133
Tensile stretch, 36
Tether force, 59
Texture correlation, 36
Thick-walled tubes, 147
Three-dimensional, 118, 146
Tissue engineering, 108, 118
Transport mechanisms, 54
Two pathway model, 77
Tyrosine kinases, 88

V
Vascular disease, 134, 146, 165
Velocity, 27, 51, 133, 146, 159
Vinculin, 85
Vortex, 137

W
Wall shear stress (WSS), 52, 133, 157
Wall thickening, 157
Wall thickness, 147, 159
Wave propagation, 15, 146
Waveforms, 146